INVESTIGATING ELECTRONICS

R. HIGGINS
A. J. C. MAY

VOLUME 1

Longman Scientific & Technical
Longman Group UK Limited,
Longman House, Burnt Mill, Harlow,
Essex CM20 2JE, England
and Associated Companies throughout the world.

© Longman Group UK Limited 1991

All rights reserved; no part of this publication
may be reproduced, stored in a retrieval system,
or transmitted in any form or by any means, electronic,
mechanical, photocopying, recording, or otherwise
without either the prior written permission of the Publishers or a
licence permitting restricted copying in the United Kingdom
issued by the Copyright Licensing Agency Ltd, 90 Tottenham
Court Road, London, W1P 9HE.

First published 1991

British Library Cataloguing in Publication Data
Higgins, R.
  Investigating electronics.
  1. Electronics
  I. Title II. May, A. J. C.
  537.5

ISBN 0-582-06003-6

Set in 11/15 Helvetica Regular

Produced by Longman Group (FE) Limited
Printed in Hong Kong

CONTENTS

Preface   iv

Acknowledgements   v

1   BASIC CONCEPTS AND COMPONENTS   1
Fuses; Switches; Resistors; Diodes; Capacitors;
Transistors; Fault finding; Soldering
Practical exercise: transistor tester
Self-assessment questions

2   DETECTOR CIRCUITS AND COMPONENTS   36
Light-dependent resistors; Thermistors; A level detector;
Relays; Relays in detector circuits; Two-stage detector
circuits; Silicon-controlled rectifiers; Silicon-controlled
rectifiers in detector circuits; Printed-circuit boards
Practical exercise: continuity tester
Self-assessment questions

3   AMPLIFICATION   59
Voltage amplifiers; Oscilloscopes; Terminal blocks;
Analogue microelectronics
Self-assessment questions

4   POWER SUPPLY UNITS   81
Transformers; Rectifiers; Filters; Voltage stabilisation;
Excess-current protection; Commercial voltage regulators;
Variable-output voltage regulators; Matrix board
construction; Practical exercise: TTL power supply unit;
Connection to mains supply; Testing insulation
Self-assessment questions

Index   120

PREFACE

This book is intended for readers wishing to be introduced to practical electronics. Little or no previous electrical or electronic knowledge is assumed and it provides a basis on which the reader can build up a basic knowledge of electronics slowly and steadily from first principles, using the many exercises.

The book is divided into four sections, namely:

- basic circuits and components,
- power supplies,
- analogue electronics, and
- power electronics.

Lists of components needed to carry out the practical exercises are also included.

The two books *Investigating Electronics* and *Investigating Digital Electronics* give an adequate cover of the EITB TR21 basic training requirement in electronics and groups who may find it useful include those studying 'Design and Technology' in schools and students taking BTEC Engineering courses having a practical electronics content designed to meet the 'common skills' element of their course.

The authors would like to thank the publishers for their friendly co-operation and helpful advice in preparing this book. They would also like to add a word of thanks to their respective wives, Yvonne and Joan, for their patience, help and encouragement during the preparation of this book.

Roy Higgins
Tony May
1990

ACKNOWLEDGEMENTS

The authors would like to acknowledge the following help received: Mr G. Tomlinson, who inspired and encouraged work on this book; Mr J. Fielding, who influenced the training methods used by one of the authors and Mr J. Hargreaves for permission to use the facilities of Training 2000 to develop the ideas put forward in the book; also the staff and students of Training 2000 who participated in the production of this book.

We are indebted to the following for permission to use copyright material:

Butterworth & Co. (Publishing) Ltd for diagrams 102 and 103 from *Modern Electrical Installation for Craft Students Second Edition* by B. Scadden (see our pages 112–14); Butterworth Heinemann for Figs. 2, 3 and 4 from *Op-Amps-Their Principles and Applications* by J.B. Dance, published by Newnes Technical Books (see our page 74); Engineering Industry Training Board for four figs. from *FYT Booklet No. 5, Electrical and Electronic Techniques* (see our pages 68–9), seven figs. from *FYT Booklet No. 12 Electronic Engineering* (see our pages 52, 53 and 63), three figs. from *ITB Instruction Module G.5 Electronic Equipment Wiring and Assembly 1* (see our pages 21, 22 and 105), and three figs. from *ITB Instruction Manual Module J.4 Electronic Maintenance 1* (see our pages 64, 110 and 111); Stanley Thornes Publishers for Figs. 1, 5, 8, 9, 15, 22 and 59 from *Electronics* by R.A. Sparkes (see our pages 1, 2, 10, 15, 16, 17 and 21).

Although every effort has been made we are unable to trace the copyright holders of *Basic Electricity Part 1* edited by J.M. Chapman and *Basic Electronics* edited by

J.M. Chapman, both of which were published by the Gower Publishing Group (see our pages 2, 3, 12, 14, 41, 82, 90 and 90–91). We would appreciate any information which would enable us to do so.

CHAPTER 1
BASIC CONCEPTS AND COMPONENTS

▶ The diagram shows a miniature light bulb. To make the bulb glow, an *electrical current* is passed through it.

▶ Water flows around a complete pipework circuit providing there is a driving force such as a pump. *Direct current* flows around a complete electrical circuit of copper wire providing there is an electrical driving force such as a battery.

▶ If the pipework is constricted the water flow will be reduced even though the pressure is unchanged. This is due to the *resistance* to water flow created by the reduced bore of the pipe.

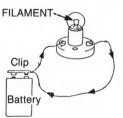

▶ Inside the glass encapsulation of the light bulb is a very thin wire called a *filament*. The filament slows down the electric current flow in the circuit as did the constriction in the water pipe, even though the battery pressure remains constant. In addition, the filament becomes white hot due to the friction of the current being forced through the filament. This is seen as light.

BASIC CONCEPTS AND COMPONENTS 1

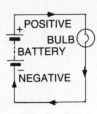

▶ To depict the layout of electrical circuits, *circuit diagrams* are used. In these, symbols are used to show components instead of pictures. The '+' in the diagram shows the *positive terminal* of the battery and the '−' the *negative terminal*.

▶ The bulb is marked '12 V, 2.2 W', where V is the *voltage* in volts and W is the power in *watts*. This indicates that when connected to a 12 V battery, a light output of 2.2 W is achieved.

$$I = \frac{2.2}{12} = 0.183 \text{ A}$$

▶ The current flowing in the circuit can be calculated by dividing the wattage by the voltage, that is

$$\text{current (A)} = \frac{\text{watts (W)}}{\text{volts (V)}}$$

In many electronic calculations, the *ampere* (A) is too large a unit and *milliamperes* (mA) are used where

$$1 \text{ A} = 1000 \text{ mA}$$
thus $\quad 0.183 \text{ A} = 183 \text{ mA}$

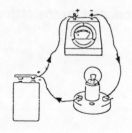

▶ The circuit current can be measured using an *ammeter*. The current must flow through the ammeter for the meter to register so the circuit must be broken to insert the ammeter. This is called connecting in series. The ammeter will read the same current as that which flows through the bulb.

• •

E X E R C I S E S

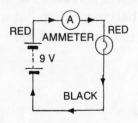

▶ Build the circuit shown in the circuit diagram. (Use black-covered wire for the return lead from the bulb to the battery. The other two leads should be red-covered wire.)

Calculate the current in milliamperes flowing in the circuit.

Fill in the missing words (1)

If the .... is connected in ....
with the light bulb there is only one path for the
.... to take. This is from the ....
of the battery through the .... and then
through the .... and back to the
.... of the battery. This means that the
.... must read the same ....
as that being taken by the ....

▶ When measuring the voltage of a battery, the *voltmeter* must be connected *in parallel* with the battery. This can be achieved by removing the ammeter and connecting the voltmeter as shown, across the battery terminals.

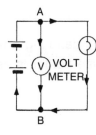

▶ The equivalent circuit diagram is as shown. Current flows from the positive terminal of the battery lighting the bulb as before, but at point A, the current separates and a very small amount of the current flows through the voltmeter. The amount of deflection shown on the voltmeter (called the voltage reading) is proportional to the battery voltage. The two currents reunite at point B and flow back to the negative terminal of the battery.

EXERCISES

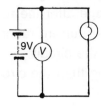

▶ Build the circuit shown in the diagram. (Use red-covered leads from the + terminal to the voltmeter and bulb and black-covered leads elsewhere.)

Note the battery voltage and the voltage across the bulb.

Fill in the missing words (2)

The driving .... given out by the battery can
be measured using a .... meter. The
.... meter must be connected in
.... with the .... and the battery.
A very small .... will flow through the

BASIC CONCEPTS AND COMPONENTS

.... meter giving a .... on the meter proportional to the battery ....

Fuses

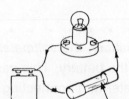

▶ A *fuse* is a piece of wire designed to melt when excessive current flows around the circuit. The size of fuse wire is selected so that it melts before damage occurs in any other part of the circuit. It is the *weakest link* in the circuit.

▶ A *cartridge fuse* is a fuse wire encapsulated in either a glass or a ceramic tube with metal end caps for making connections to the fuse holder.

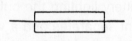

▶ The British Standard symbol for a fuse denoting a fuse wire passing through a cartridge or fuse holder is as shown.

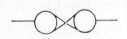

▶ Another widely used symbol for the fuse is shown in the diagram.

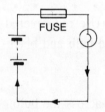

▶ The fuse should always be placed in the positive or live lead of the circuit. This ensures that the wiring and components are at a low or zero voltage after the fuse has melted. The fuse should have a higher current rating than the normal circuit current dictated by the circuit components, but a lower current rating than the circuit cable.

E X E R C I S E S

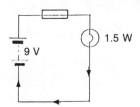

▶ Calculate a suitable fuse size for the circuit shown.

▶ Build the circuit shown, having the fuse in the positive lead.

▶ Fill in the missing words (3).

A fuse is the .... link in a circuit and is designed to .... if excessive .... flows in the circuit.

Switches

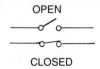

▶ Switches come in a wide variety of shapes and sizes. The simplest of these is the *single-pole single-throw* (SPST) or on/off switch.

The symbols for the SPST switch in open and closed positions are as shown. It has only two connections and pivots at the input end to close or open a circuit.

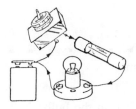

▶ An on/off switch is used to isolate a particular circuit so it is the first component in the circuit. The terminal in the middle of the body of the switch is the common or input terminal and is the pivot point of the switch.

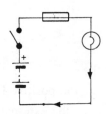

▶ A circuit diagram of a circuit having an SPST switch is as shown.

BASIC CONCEPTS AND COMPONENTS

EXERCISES

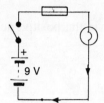

▶ Build a circuit as shown in the diagram.

Operate the switch and observe the effect on the bulb.

Fill in the missing words (4).

If a circuit has to be ...., a switch must be wired into the circuit. An SPST switch is the abbreviation of .... and is simply an .... switch.

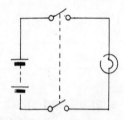

▶ A DPST switch is a *double-pole single-throw switch*. It can be regarded as two single-pole single-throw switches in one case as it contains two electrically separate single-poles switching simultaneously. The broken line denotes simultaneous switching. It can be used for totally isolating the external circuit from the supply.

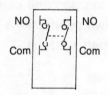

▶ As the DPST switch contains two SPST switches it has four connections. The pivot point of each switch is called the common. The other connection of each switch is called the normally open.

EXERCISES

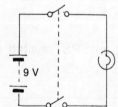

▶ Build a circuit as shown in the diagram.

Operate the switch and observe the *effect* on the bulb.

Fill in the missing words (5).

The DPST switch appears to behave like the .... switch but rather than just .... the positive bulb circuit connection, it totally .... the external circuit from the supply.

▶ *A single-pole double-throw switch* (SPDT) is a single switch which changes from one circuit to another with no 'off' position. It is called a *change-over* switch and has three connections, known as (i) normally open, (ii) normally closed and (iii) common. The common connection is used to join an incoming lead to either the normally open or normally closed connection.

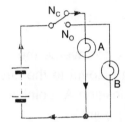

▶ In the circuit diagram, lamp A will be lit as it is connected to the normally closed connection of the switch.

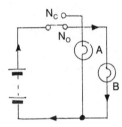

▶ When the switch is operated, lamp A will go out and lamp B will light as the normally open contact closes.

• •

E X E R C I S E S ▶ Build the circuit shown.

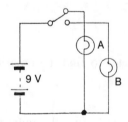

Note which bulb lights in the NC position.

Note which bulb lights in the NO position.

Will both bulbs switch 'off' simultaneously?

Will both bulbs switch 'on' simultaneously?

Fill in the missing words (6).

The SPDT is a .... over switch first supplying current to .... and then .... without any .... position.

• •

BASIC CONCEPTS AND COMPONENTS

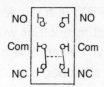

▶ The *double-pole double-throw switch* (DPDT) has two change-over switches in one case. It changes both switches simultaneously when operated.

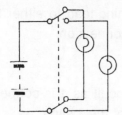

▶ The circuit shows that two totally separate circuits can be selected. No off position is available. When one circuit is selected, the other is isolated.

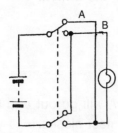

▶ The DPDT switch is quite often used to reverse the polarity of a point in the circuit with respect to the circuit supply voltage. With the switch in position A, point A will be a positive voltage with respect to point B.

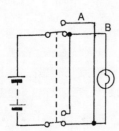

▶ With the switch in the B position point A will be a negative voltage with respect to point B.

E X E R C I S E S ▶ Build the circuit shown.

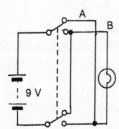

With switch in position A use a voltmeter to find the polarity of A with respect to B.

With switch in position B use a voltmeter to find the polarity of A with respect to B.

Fill in the missing words (7).

A .... switch is quite often used to .... the supply to a circuit or to .... one circuit while .... in another circuit.

Resistors

▶ The component shown is called a *resistor*. Resistors can be made from carbon and resist or slow down the current flow in a circuit.

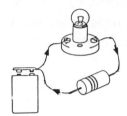

▶ When a resistor is connected in series with a light bulb, the light bulb current will be reduced and consequently the brightness of the light bulb will be reduced.

▶ Resistors come in a wide range of resistance values, each one with its own unique *colour code*, usually bands or rings of colour placed round the resistor.

```
BLACK   = 0
BROWN   = 1
RED     = 2
ORANGE  = 3
YELLOW  = 4
GREEN   = 5
BLUE    = 6
VIOLET  = 7
GREY    = 8
WHITE   = 9
```

▶ The first band determines the first digit, the second band determines the second digit and the third band determines the number of zeros to be added to these digits.

With reference to the diagram, colour bands of yellow/violet/red indicate 4/7/2 zeros, that is, a resistance of 4700 ohms.

• •

EXERCISES

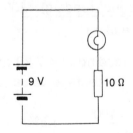

▶ Using the resistor colour code, select resistors of 10 Ω, 47 Ω and 100 Ω. Build the circuit shown and note the brightness of the lamp.

Replace the 10 Ω resistor with the 47 Ω resistor and note the brightness of the lamp.

Replace the 47 Ω resistor with the 100 Ω resistor and note the brightness of the lamp.

State the effect of increasing the resistance in a circuit.

Fill in the missing words (8).

BASIC CONCEPTS AND COMPONENTS

Resistors are often made from .... and as they are made in a variety of values a .... is used to recognise them. If a resistor is placed in .... with a light bulb the .... through the light bulb will be .... and the light given out will be ....

• •

▶ The value of a resistor can be measured by using an *ohmmeter*. Only the ohmmeter and resistor are used and so the ohmmeter requires an internal battery.

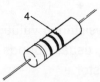

▶ When measuring the resistance of a resistor it may be found that the value shown on the ohmmeter does not exactly correspond to that given by the colour bands. This is because resistors are manufactured not exactly to a precise value but with a *tolerance*. The value of tolerance is denoted by a fourth band on the resistor.

GOLD ± 5%

SILVER ± 10%

▶ The most common colours for the fourth band are gold or silver although others are used. The gold band means ± 5% tolerance. This means that the resistor could be as high as 5% higher than the colour bands suggest or as low as 5% lower than the colour bands suggest. A silver band means ± 10% tolerance.

The percentage value is calculated from:

$$\frac{(\text{ohmmeter reading} - \text{'band' reading}) \times 100}{\text{ohmmeter reading}}$$

Thus for a resistor having a band reading of 4.7 kΩ and the ohmmeter indicating 5 kΩ, the percentage value is

$$\frac{(5.0 - 4.7) \times 100}{5.0} = \frac{0.3 \times 100}{5.0} = +6\%$$

INVESTIGATING ELECTRONICS

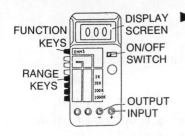

▶ When using a *digital multimeter* as an ohmmeter, switch on the meter and connect the positive and negative leads to the meter.

The function key marked 'ohms' must now be pressed.

The range keys give a selection of resistances, often the ranges being 2 k–20 k, 200 k–2000 k. The letter k means kilo or 1000 so that 2 k means 2000 ohms. Choosing any range and shorting the leads together will give a display of 000 which means no resistance or 0 ohms.

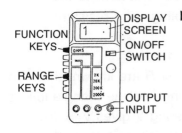

▶ It is necessary to choose the correct range key to suit the resistor to be measured. If a resistor of 10 k is to be measured and too low a range is selected, e.g. 2 k range, a 1 will appear at the far left of the display. This indicates that the range is too low.

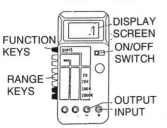

▶ If too high a range has been chosen the display will read low 1.0 if the 200 k range is chosen or .1 if the 2000 k range is chosen.

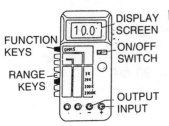

▶ The correct choice of range to measure a 10 kΩ resistor would be the 20 k range.

• •

E X E R C I S E S ▶ Connect a 10 Ω resistor to the ohmmeter. Note the reading and determine the percentage tolerance value.

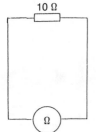

Connect a 47 Ω resistor to the ohmmeter. Note the reading and determine the percentage tolerance value.

Connect a 100 Ω resistor to the ohmmeter. Note the reading and determine the percentage tolerance value.

BASIC CONCEPTS AND COMPONENTS

Fill in the missing words (9).

The resistance value of a resistor can be checked using an .... The .... has an internal .... because no external power supply is used. If the .... reading is different from the ...., this suggests the difference is due to the manufacturer's tolerance.

- -

▶ A triangle as shown may be used to determine the relationship between current, voltage and resistance.

▶ Provided two of the values *V*, *I*, or *R* are known, the third value can be calculated. Covering *I* on the triangle leaves V/R, so *I* = V/R. Similarly, covering *V* gives *V* = IR. Finally, covering *R* gives *R* = V/I. These relationships are known as *Ohm's law*.

- -

E X A M P L E S

▶ *Example 1*
Calculate the circuit current in mA.

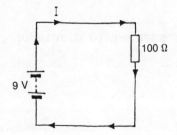

$$I = \frac{V}{R}, \text{ i.e. } I = \frac{9}{100} = 0.09 \text{ A}$$

$$0.09 \text{ A} = 0.09 \times 10^3 = 90 \text{ mA}$$

▶ *Example 2*
Calculate the battery voltage.

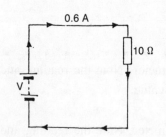

$$V = I \times R = 0.6 \times 10 = 6 \text{ V}$$

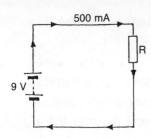

▶ *Example 3*
Calculate the circuit resistance.

$$500 \text{ mA} = 500 \times 10^{-3} = 0.5 \text{ A}$$

$$R = \frac{V}{I} = \frac{9}{0.5} = 18 \text{ }\Omega$$

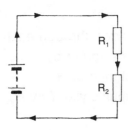

▶ Connecting two or more components so that only one current path exists is called *series* connection.

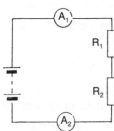

▶ When ammeters are connected in different parts of a series-connected circuit, they will both have the same reading. The *same current* flows in all parts of a series-connected circuit. Thus $A_1 = A_2$.

When calculating the current in a series circuit $I = V \div R_T$ where R_T is the total resistance of the circuit.

The total resistance of a series circuit can be found by adding all the resistance values together:

$$\text{e.g.} \quad R_T = R_1 + R_2$$

EXERCISES

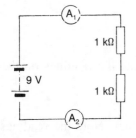

▶ Build the circuit shown in the diagram using two 1 kΩ resistors and a 9 V battery. Note the ammeter readings.

Change both the resistors to 10 kΩ. Note the ammeter readings.

Calculate the circuit current by using Ohm's law.

Fill in the missing words (10).

When .... flows around a ....
circuit it can neither .... or decrease but
must remain the .... Two ....

BASIC CONCEPTS AND COMPONENTS

placed anywhere in the circuit will therefore read the
....

▶ Every component in a circuit will have a measurable voltage across it. If two or more components are connected in series the *voltage drops* (V_1 and V_2) will total the supply or battery voltage (V) when added together.

▶ In a series circuit the same current flows through each component so voltage drop (V_1) will be found by multiplying the current I by the resistance of R_1 ($V_1 = I \times R_1$). V_2 can be found by multiplying I by R_2 ($V_2 = I \times R_2$).

To find the supply voltage the current I must be multiplied by the total resistance of the circuit ($V = I \times R_T$).

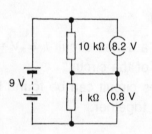

▶ Since $V = I \times R$ and since the current is constant due to the resistors being connected in series, the highest resistance will have the highest voltage drop and the lowest resistance the lowest voltage drop.

• •

EXERCISES

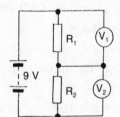

▶ Build the circuit shown in the diagram using two 1 kΩ resistors and a 9 V supply.

Note the reading on voltmeters.

Change R_1 for a 10 kΩ resistor and note the readings on the voltmeters.

Reverse the positions of the two resistors. Note the readings on the voltmeters.

Check that the voltage drops across the readings are roughly equal to the supply voltage.

Calculate the values of V_1 and V_2 in each case using Ohm's law.

Fill in the missing words (11).

The .... drops in a .... circuit .... together will total the .... The .... resistance will have the most .... and the .... resistance will have the least ....

▶ The resistors described so far are known as fixed resistors. This picture shows a *variable resistor*, it has a shaft to which a knob is attached for varying the resistance value and three terminals shown as X, Y, Z.

▶ If the X and Z terminals only are used the resistance will not vary but remain a fixed maximum value.

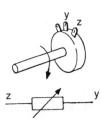

▶ If the Y and Z terminals are used the resistance will vary, increasing in value as the shaft is turned clockwise. When connected in this manner the variable resistor is called a *rheostat*. The symbol for a rheostat is shown at the bottom of the diagram.

▶ An alternative symbol for the rheostat. Y and Z show the terminals used. The terminal Y is termed the *wiper* and this is the moving terminal of the rheostat.

▶ When the terminals X and Y are used the resistance will be increased by turning the shaft anti-clockwise.

BASIC CONCEPTS AND COMPONENTS

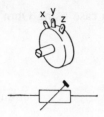

▶ A *preset resistor* is constructed exactly like the variable resistor, but the shaft is reduced in length or even eliminated and instead of a knob to control the variable resistance, a screwdriver slot is used.

Once the resistance is set with a screwdriver it is said to be preset and further adjustment is not usually necessary. The symbol for the preset resistor is as shown at the bottom of the diagram.

▶ Some preset resistors are soldered to printed-circuit boards. To facilitate this, the pins protrude from the back of the resistor. These are known as *button preset resistors*.

▶ When preset resistors are not encased, they are known as *skeleton preset resistors*, as the internal working can be seen.

EXERCISES

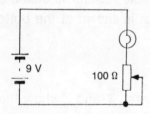

▶ Build the circuit shown using the X and Y terminals on the rheostat.

Turn the shaft clockwise and note the effect on the light emitted by the lamp.

Reconnect the rheostat using the Y and Z terminals.

Turn the shaft clockwise and note the effect on the light emitted by the lamp.

Fill in the missing words (12).

When the rheostat is turned to a high value of ...., the current will be .... and the bulb will .... When the rheostat is turned to a low value of ...., the current will be .... and the bulb will ....

16   INVESTIGATING ELECTRONICS

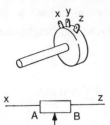

▶ When all three terminals are used the variable resistor becomes a *potentiometer*.

The symbol for a potentiometer is as shown at the bottom of the diagram.

With a voltage of $V+$ at X and zero volts at Z, the voltage at Y will change progressively from $V+$ when Y is at A to zero when Y is at B. With Y at the position shown the voltage at Y is about $(V+)/2$.

EXERCISES

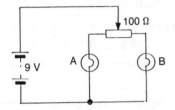

▶ Build the circuit shown in the diagram.

Turn the shaft of the potentiometer anti-clockwise and note the effects on the light emitted by the lamps.

Turn the shaft of the potentiometer clockwise and note the effects on the light emitted by the lamps.

Fill in the missing words (13).

When the wiper is moved to the left of the track the
.... between the battery and the bulb A is
.... and the current is ....,
making the bulb .... This will mean that the
.... between the battery and the bulb B has
been .... and the bulb will be
....

▶ Not all variable resistors have mechanical moving parts, the picture shows a different type of variable resistor called a *thermistor*. This resistor has a high resistance when it is cold but when it is heated the resistance falls to a lower level.

The symbol for a thermistor is as shown at the bottom of the diagram. The '$-t°$' indicates that the resistance decreases with increasing temperature.

BASIC CONCEPTS AND COMPONENTS

EXERCISES ▶

Build the circuit shown in the diagram. Note the brightness of the lamp.

Heat the thermistor. Note the brightness of the lamp.

Fill in the missing words (14).

When cold, the thermistor has a .... resistance so the current .... and the bulb is .... When the thermistor is heated the resistance is .... allowing the current to .... which makes the bulb ....

▶ Another variable resistor is the *light dependent resistor* (LDR). This resistor has a high resistance in poor light and a low resistance in strong light.

The symbol for an LDR is as shown at the bottom of the diagram. The direction of the arrows shows that it accepts light. LDRs are used as light detectors.

EXERCISES ▶

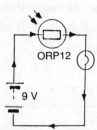

Build the circuit shown in the diagram. Note the brightness of the lamp with the LDR in a dim light.

Shine a strong light on the LDR. Note the brightness of the lamp.

Fill in the missing words (15).

If the LDR is in poor light its resistance is .... making the current .... and the bulb .... When in good light, the resistance of the LDR .... so that the current .... and the bulb ....

18   INVESTIGATING ELECTRONICS

Diodes

▶ *Junction diodes* vary a great deal in appearance. The one shown is a black plastic tube with a silver line at one end. The line denotes the output or *cathode* end of the diode. The other end of the diode is called the *anode* end.

▶ A diode only passes current from the anode end to the cathode end. When the anode end is at a higher voltage than the cathode end, current flows and the diode is *forward biased*. Virtually no current flows when the cathode end is at a higher voltage than the anode end and the diode is then *reverse biased*. The diode symbol is shown at the bottom of the diagram.

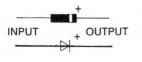

▶ Sometimes on diagrams, a plus sign is shown at the cathode end of the picture or symbol. This sign denotes the cathode end of the diode and has nothing to do with the polarity of the supply.

- -

E X E R C I S E S

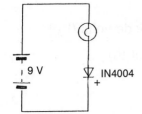

▶ Build the circuit shown in the diagram. Note the brightness of the lamp.

Rewire the diode so that the cathode end faces the positive terminal of the battery. Note the brightness of the lamp.

Fill in the missing words (16).

A diode has .... resistance when connected with .... biasing but ....
resistance when connected with .... biasing.
Current will flow through the diode when the ....
is most positive.

- -

BASIC CONCEPTS AND COMPONENTS

The light-emitting diode or LED is a diode which releases energy as light making it useful as an indicator.

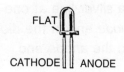

▶ The flat portion on the rim of the plastic case denotes the cathode lead of the diode. The most common colours are Red, Green and Yellow.

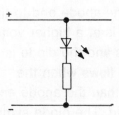

▶ The LED is an extremely low-current device and can be easily damaged. For this reason it should always have a *current-limiting resistor* connected in series with it. Note the diode symbol. The direction of the two arrows shows that the device gives off light.

$$R = \frac{V_s - V_F}{I_F}$$

▶ A current-limiting resistor must be large enough to protect an LED and small enough to allow sufficient voltage across the diode for it to operate. Typically the operating current is 10 mA (0.01 A) at a voltage of 2 V. Thus, for a 9 V supply, the voltage drop across the limiting resistor is 7 V and since *resistance is the ratio of voltage to current*, its resistance, R, is:

$$7 \text{ (V)} \div 0.01 \text{ (A)} = 700 \text{ } \Omega$$

In general,

$$R = \frac{[\text{battery volts } (V_s) - \text{diode volts } (V_F)]}{\text{diode current } (I_F)}$$

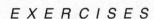

EXERCISES ▶ Build the circuit shown in the diagram. Note whether the LED emits light.

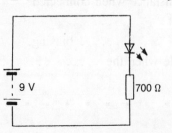

Connect the LED so that its cathode end is towards the + terminal of the battery. Note whether the LED emits light.

Calculate the limiting resistor value in order to use it with a 12 V battery.

INVESTIGATING ELECTRONICS

Fill in the missing words (17).

A light-emitting diode releases .... as
.... making it useful as an ....

Capacitors

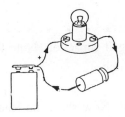

▶ When a large-value capacitor is placed in series with a lamp, current flows into the capacitor, lighting the lamp for a short period. Just as a cistern only accepts water until it is full, a capacitor only accepts current until it is charged with electrons. When the battery is removed and the leads are connected to each other (shorted out), the flow of electrons from the capacitor causes the lamp to light briefly.

▶ The circuit diagram of the capacitor test circuit is shown in the diagram.

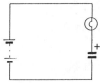

▶ There are two main types of capacitors, namely *polarised* and *unpolarised*. Unpolarised capacitors consist essentially of two metal plates separated by an insulating material and can be connected with either lead to the battery + terminal. Three types are shown in the diagram.

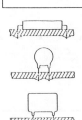

▶ The circuit symbol for an unpolarised capacitor is as shown.

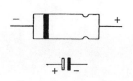

▶ The *electrolytic capacitor* is a polarised capacitor. It is sealed in a metal can by crimping the end. This end is the positive end of the capacitor and must always be connected to the most positive end of the circuit. The symbol for the polarised capacitor is as shown.

BASIC CONCEPTS AND COMPONENTS

▶ Some electrolytic capacitors have the leads at the bottom rather than at each end. This type of capacitor has the negative lead marked with an arrow or a line of minus signs.

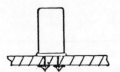

▶ This type of capacitor is soldered flat to the circuit board.

• •

E X E R C I S E S ▶ Build the circuit shown using a discharged capacitor. Note the effect on the lamp when closing the switch.

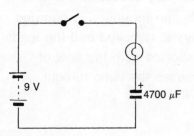

Open the switch and remove the battery. Now short the leads out which were previously connected to the battery. Note the effect on the lamp when closing the switch.

Fill in the missing words (18).

The capacitor will only accept current when it is .... As soon as it is .... the current will stop flowing and the light will be ....

• •

Transistors

▶ *Transistors* can vary a great deal in appearance, some are in black plastic packages and some are in shiny silver cans. They also come in a wide range of shapes and sizes but they all have three leads.

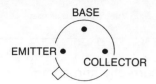

▶ It is extremely important to connect transistors in the correct way in a circuit or they will be damaged. Charts showing the position of the leads are available for every type of transistor. The diagram shows the base of a very common transistor, the BFY 51. The leads are called the *base*, the *collector* and the *emitter*. On this particular transistor the emitter is identified by a pip on the transistor case.

▶ The circuit symbol for a metal cased transistor is as shown. This is a *NPN transistor*. A reverse polarity transistor is a *PNP transistor*.

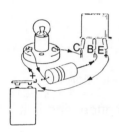

▶ When there is no base voltage, no current flows from the collector to the emitter. Hence the base voltage controls the transistor. The base must be connected to a voltage before the transistor will switch on and pass current from the collector to the emitter and hence light the lamp.

EXERCISES

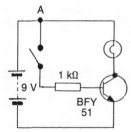

▶ Build the circuit shown. Note the effect of opening and closing the switch.

Fill in the missing words (19).

A transistor can be used as a .... A circuit is completed between the .... and the .... lighting the bulb only when a .... current is present to switch on the transistor.

BASIC CONCEPTS AND COMPONENTS 23

Fault finding

▶ Circuits may not always operate correctly and some detective work may be necessary. This can be achieved by asking questions.

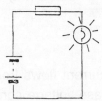

▶ What do I expect the circuit to do?

In this case the answer is that the bulb must light.

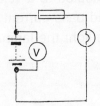

▶ If the bulb does not light the next question is:

is there a supply voltage from the battery and is it sufficiently high?

Connect a voltmeter across the battery and if the supply voltage is low change it. Upwards of 4 V should cause the bulb to glimmer at least.

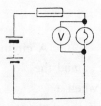

▶ If the battery voltage is satisfactory:

does the battery voltage appear across the bulb? Check by using a voltmeter.

▶ If the battery voltage does appear across the bulb then the fault is either in the bulb or the bulb holder.

Does the bulb have a resistance value?

Check the bulb with an ohmmeter. If the bulb resistance is infinity then it will need replacing.

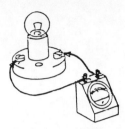

▶ If the bulb has a resistance reading:

does the bulb and holder have a resistance value when connected in the holder?

If the answer is 'no' then a fault exists in the bulb holder which may be found by inspection.

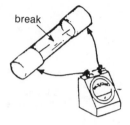

▶ If the battery voltage does not appear across the bulb then the fuse or the connecting leads must be at fault.

Does the fuse have a low resistance reading?

Check the fuse with an ohmmeter. If the fuse reads infinity then it needs replacing. (Note: fuses may have a good reason for melting. Check that the circuit is correctly wired.)

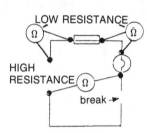

▶ If the fuse wire is intact, then test each connecting lead in turn using an ohmmeter.

Does each connecting lead have a low resistance reading?

(Note: when using an ohmmeter disconnect the battery first.)

When a fault occurs a *flow chart* can be drawn up as a guide to the probable fault location.

BASIC CONCEPTS AND COMPONENTS

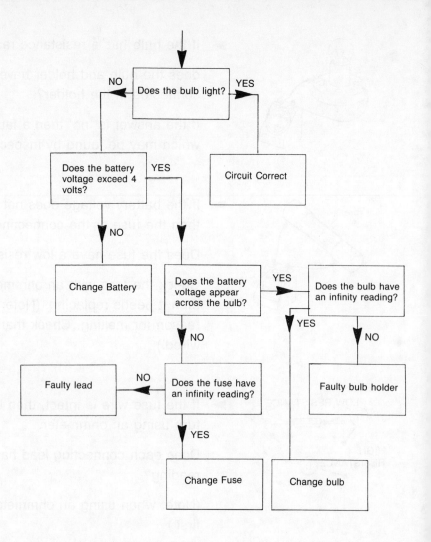

EXERCISES

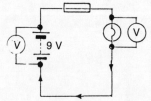

▶ Randomly mix a number of flat batteries, fuses and bulbs some of which are sound and others faulty.

Build the circuit shown in the diagram, randomly selecting from the items given above.

Use the test procedure outlined to find the defective components.

26   INVESTIGATING ELECTRONICS

Soldering

▶ Inside an electrical *soldering iron* is a heating element which becomes hot when an electrical current flows through it. The element is in contact with the part of the soldering iron called the bit and the face of the bit makes contact with the surfaces to be soldered.

▶ Although some soldering irons operate at mains voltage a much safer type of soldering iron is the low-voltage soldering iron. The 240 volt mains supply is reduced to a lower voltage by a transformer before supplying the soldering iron.

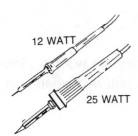

▶ The heat given off by a soldering iron is dependent on its *power rating*. The higher the wattage of the soldering iron the more heat the soldering iron will dispense.

▶ When soldering, the iron is held as when writing with a pencil.

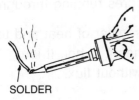

▶ Check that the soldering iron has heated up to the correct temperature by melting a little solder on to the bit. The solder should melt immediately to a shiny surface and white smoke should be given off. This is called *tinning*.

BASIC CONCEPTS AND COMPONENTS

▶ Wipe the bit on a sponge pad to clean and remove the excess solder. The bit should be evenly covered with solder. If it is not, clean the bit and tin again.

▶ The soldering iron must be kept in its stand when it is not in use. This will ensure that the hot bit does not accidentally melt the lead to the iron, which could cause electric shock or fire.

FACE

▶ *Soldering iron bits* are usually made from copper because copper is a very good conductor of heat. Some bits have an iron-plated face to the bit which reduces wear and so gives a longer service life.

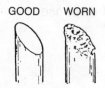

GOOD   WORN

▶ Unplated bits may be filed when they are worn but iron-plated bits must *never* be filed but renewed when worn.

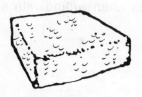

▶ Bits must be cleaned frequently when they are hot by rubbing them on a damp sponge. A dirty bit will produce poor-quality soldering.

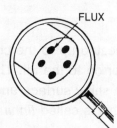

FLUX

▶ *Soft solder* is an alloy of tin and lead and the type used for electronic work has *resin flux* cores running through it.

The flux is used to speed up the transfer of heat and to prevent oxidisation of the metal when it is hot. It is impossible to solder satisfactorily without flux.

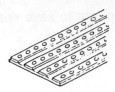
▶ *Stripboard* is a very commonly used circuit board for the assembly of one-off circuits. It consists of resin-bonded paper board with copper strips running from end to end. A series of holes accommodate the components.

▶ Mount the components on the non-strip side of the circuit board to form the circuit pattern required.

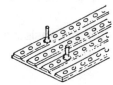
▶ Invert the board and solder the component leads to the copper strips.

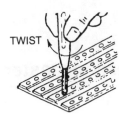
▶ Trim off the excess component leads and if necessary break the tracks using a *stripboard cutter*. This may be necessary if unwanted connections have been made or components are shorted out by the tracks. Finally solder any supply leads to the board.

▶ Some components are easily damaged by heat from a soldering iron.
   It is advisable to clamp a *heat shunt* to the leg of the component between the component and the joint being soldered. This will shunt the heat away from the component because of its large metal mass, thus keeping the component cool.

▶ Resistors, capacitors and diodes should be mounted flush with the board.

▶ Alternatively they can be mounted upright to save space.

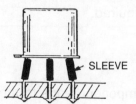
SLEEVE

▶ Transistors are best mounted above the board so that the increased lead length reduces possible damage to the transistor by overheating.

It is advisable to sleeve the leads to prevent short circuiting.

Always use a heat shunt when soldering transistors and diodes (see above).

Practical exercise: A transistor tester

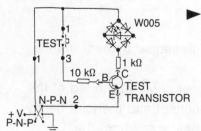

▶ The circuit diagram is as shown.

▶ Prepare the stripboard.

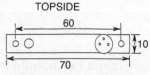
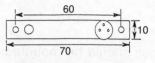
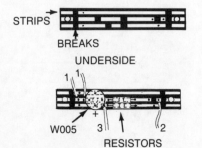

▶ Cut a piece of stripboard along a line of holes leaving three clear strips.

▶ File the edges smooth.

▶ Mount and solder the topside components.

▶ Cut the strips in the relevant positions with a strip cutter.

▶ Mount and solder the underside components.

▶ Trim off the excess leads leaving the top smooth for mounting to the lid.

▶ Fit a piece of insulation tape between the LED and the test socket on the top surface to cover the lead ends.

Build the transistor tester.

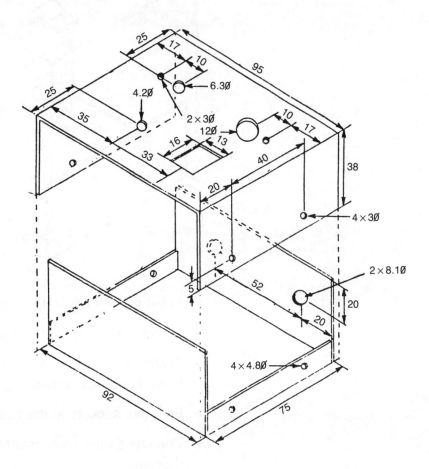

▶ Cut the base from a suitable material such as an aluminium alloy and drill the holes while flat. The 4.8 mm fixing holes can be left.

▶ Fold the base into shape. (Dimensions on diagram are in mm.)

▶ Cut the lid and drill all the holes.

▶ Fold the lid and place over the base. Mark the 4.8 mm fixing holes through the lid fixing holes and drill.

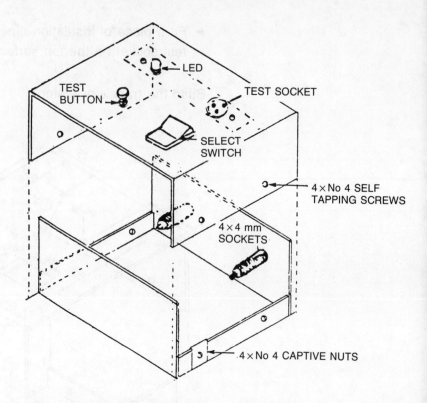

► Label the top cover and cut out the holes.

► Mount the test button and select switch.

► Fasten the circuit board into position using 3 mm instrument head screws.

► Fit 4 mm sockets to the base, one black and one red.

► Connect all the components as shown in the circuit diagram.

► Fit the lid using No 4 captive nuts and No 4 self-tapping screws.

COMPONENT LIST

1   10 kΩ RESISTOR
1   1 kΩ RESISTOR
1   SUB-MINIATURE PUSH BUTTON (NO)
1   TRANSISTOR HOLDER
1   MINIATURE ROCKER SWITCH (DPDT)
1   STANDARD LED RED

1 BRIDGE RECTIFIER
1 4 mm SOCKET (RED)
1 4 mm SOCKET (BLACK)
4 RUBBER FEET
4 CAPTIVE NUTS
4 SELF-TAPPING SCREWS
2 3 mm SCREWS
2 3 mm NUTS
2 3 mm WASHERS
1 STRIPBOARD

Self-assessment questions

1. Soldering iron bits are made from copper because it is:

(a) a good conductor of electricity;
(b) a good conductor of heat;
(c) a very cheap metal;
(d) a non-ferrous metal.

2. Solder is an alloy of:

(a) tin and zinc;
(b) zinc and lead;
(c) tin and lead;
(d) tin and aluminium

3. An LED is:

(a) a light bulb;
(b) a photo cell;
(c) a light-dependent resistor;
(d) a light-giving diode.

4. Resistance is measured with a:

(a) voltmeter;
(b) wattmeter;
(c) ammeter;
(d) ohmmeter.

5. Conventional d.c. current flows:

(a) from negative to positive;
(b) from positive to negative;
(c) in both directions;
(d) does not flow at all.

6. An ammeter should be connected in:

(a) series with the circuit;
(b) parallel with the circuit;
(c) parallel with part of the circuit;
(d) all of these.

7. The fourth band on a resistor is:

(a) the last number;
(b) the number of zeros;
(c) the tolerance;
(d) the working voltage.

8. A diode passes current from:

(a) anode to cathode;
(b) cathode to anode;
(c) both ways;
(d) blocks current in either direction.

9. A transistor has three leads:

(a) emitter/cathode/anode;
(b) collector/emitter/anode;
(c) collector/emitter/base;
(d) cathode/collector/base.

10. Resistors are often made from:

(a) copper;
(b) aluminium;
(c) carbon;
(d) silicon.

Answers to self-assessment questions

1. (b); 2. (c); 3. (d); 4. (d); 5. (b); 6. (a); 7. (c);
8. (a); 9. (c); 10. (c).

Missing words

(1) Ammeter, series, current, positive terminal, ammeter, bulb, negative terminal, ammeter, current, bulb.
(2) Pressure, volt, volt, parallel, bulb, current, volt, reading, voltage.
(3) Weakest, melt, current.
(4) Broken, single-pole single-throw, on/off.
(5) SPST, isolating, isolates.
(6) Change, circuit A, circuit B, off.
(7) DPDT, reverse, isolate, switching.
(8) Carbon, colour code, series, current, reduced, reduced.
(9) Ohmmeter, meter, battery, meter, colour code.
(10) Current, series, increase, same, ammeters, same.
(11) Voltage, series, added, supply voltage, highest voltage drop, smallest voltage drop.
(12) Resistance, low, dim, resistance, increased, bright.
(13) Resistance, reduced, increased, bright, resistance, increased, dim.
(14) High, is low, dim, reduced, rise, bright.
(15) High, low, dim, reduces, increases, brightens.
(16) Low, forward, high, reverse, anode.
(17) Energy, light, indicator.
(18) Discharged, charged, off.
(19) Switch, collector, emitter, base.

CHAPTER 2
DETECTOR CIRCUITS AND COMPONENTS

Light-dependent resistors

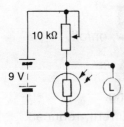

▶ A variable resistor and *light-dependent resistor* are connected as shown to form a voltage divider circuit. In strong light, the LDR has a low resistance of about 10 Ω, so the output voltage is low.

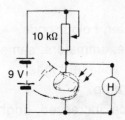

▶ In dim light, the resistance of the LDR increases to about 1 MΩ, thus the output voltage is now near to the supply voltage. When the output voltage is used to, say, switch on a light, then the circuit is a *darkness detector*.

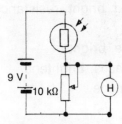

▶ The LDR and variable resistors can be interchanged and instead of detecting the amount of darkness, it will detect the amount of light. The output voltage is high as the resistance of the LDR is low, giving a high voltage output in strong light.

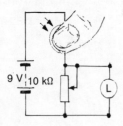

▶ If we increase the LDR resistance by placing it in the dark, the voltage across the variable resistor falls drastically giving a low voltage output to show that no light is present. The variable resistor is used to set the sensitivity of the circuit.

EXERCISES ▶ Build the circuit shown in the diagram.

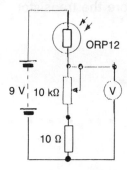

Set the variable resistor to give a high resistance value and measure the output voltage.

Cover the LDR and measure the output voltage.

Adjust the variable resistor until the output voltage is low in daylight.

Slowly increase the resistance of the variable resistor until the output reverts back to a high voltage. The circuit is now sensitive, as the slightest shade will lower the output voltage.

Fill in the missing words (1).

The .... resistor sets the .... of the circuit. When used as a light detector, the variable resistor must be set to a .... resistance.

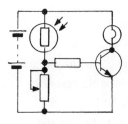

▶ A light-sensitive voltage divider can be used to control a transistor by connecting the transistor base to its output. The advantage of using a transistor in a *light-detecting circuit* is to create a greater degree of *sensitivity*.

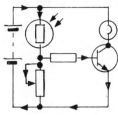

▶ When the LDR is in good light, its resistance is low and the transistor passes current. Some current flows through the variable resistor and some into the transistor base to switch on the transistor.

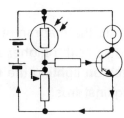

▶ When the variable resistor is set to a high resistance value the base current will be the greater, switching on the transistor at relatively low light levels.

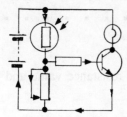

▶ When the variable resistor is set to a low resistance value the majority of the current flows away from the transistor and a high light level is needed before the transistor switches on.

• •

E X E R C I S E S

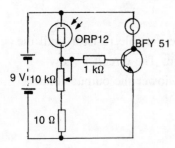

▶ Build the circuit shown in the diagram.

With the LDR in full light, adjust the variable resistor until the light goes out.

Increase the resistance of the variable resistor until the light comes on. At this setting the slightest shade will put out the light.

Change the components necessary to change the light detector to a dark detector.

Fill in the missing words (2).

The light detector has been made more .... to light variation due to the addition of a .... The operation point of the .... can be set using a .... When the variable resistor is set to a low value of resistance, very little .... current flows even in light and the .... does not switch on.

• •

Thermistors

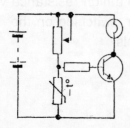

▶ When a *thermistor* is used as shown in the diagram a *cold detector* results. The colder the thermistor the higher is its resistance so the transistor will switch on lighting the bulb. The circuit is suitable for warning against frost.

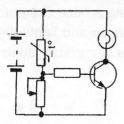

▶ If the positions of the thermistor and variable resistor are reversed, the circuit becomes a *heat detector* which can warn of the presence of a hot surface such as a ceramic hob on a cooker. The circuit is also suitable for warning against defrosting of refrigerators and freezers.

E X E R C I S E S

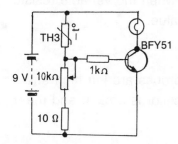

▶ Build the circuit shown in the diagram.

Adjust the variable resistance until the light goes off.

Heat the thermistor and note the effect on the light.

Change the circuit components to produce a cold detector.

Fill in the missing words (3).

The temperature detector circuit can act as a .... surface warning device or a .... warning device. This could make it suitable for use in appliances such as .... and ....

A level detector

▶ A similar basic detector circuit can be used to indicate the level of fluids, such as tap water, which conduct electricity.

DETECTOR CIRCUITS AND COMPONENTS

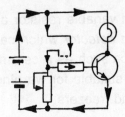

▶ Probes dropped in the water complete a base circuit to the transistor. The transistor switches on and lights the bulb provided the variable resistance is correctly adjusted.

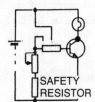

▶ Because tap water has little resistance, a safety resistor is connected in series with the variable resistor to prevent the battery being short circuited when the variable resistor is set on a very low resistance value.

▶ When the fluid level drops, the probes are left high and dry. The base circuit to the transistor is broken and the light is out.

The circuit can also be used as a *continuity tester*.

E X E R C I S E S ▶ Build the circuit shown in the diagram.

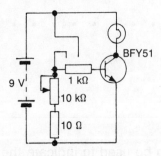

Set the variable resistor to give maximum sensitivity with the probes shorted together.

Drop the probes into a cup of water and note the effect on the light.

Remove one probe from the water and note the effect on the light.

Use the circuit to test the continuity of a conductor.
Fill in the missing words (4).

It is sometimes undesirable for .... levels to fall too low. The .... circuit can be utilised to .... a lack of conductive .... The .... of cables can also be checked with this circuit.

Relays

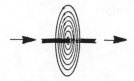

▶ When a conductor is carrying a current, a *magnetic field* is generated which encircles the conductor. The size of this magnetic field is determined by the amount of current flowing in the conductor.

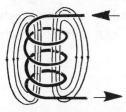

▶ Winding the conductor to form a coil causes the magnetic fields produced around each turn to intertwine to produce one large magnetic field around the coil.

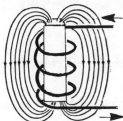

▶ The magnetic field can be further strengthened by inserting an iron core in the centre of the coil.

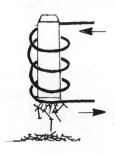

▶ As long as the current flows in the coil, the iron core becomes *an electromagnet* and attracts metal objects.

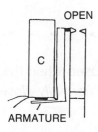

▶ The *armature* of a relay is a metal flap which is attracted by an electromagnet. It has a dog-leg-shaped arm which pushes a contact at the top of the relay. When the coil is not carrying current the armature does not move and the contacts remain open. The contact position is called normally open (N_O).

DETECTOR CIRCUITS AND COMPONENTS

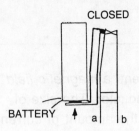

▶ When the coil is connected to a supply voltage, the coil will energise and attract the armature. This pushes the contacts together and completes a circuit between points a and b. Such a device is called a *relay*.

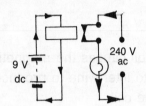

▶ A relay is useful for using a low-voltage circuit to operate a high-voltage circuit. For example, the low-voltage coil can be energised from a battery and the contacts, which are high current and voltage rated, can control a higher-voltage circuit.

E X E R C I S E S ▶ Build the circuit shown in the diagram.

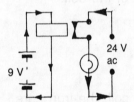

Connect the battery and switch on. Note the effect on the lamp.

Switch off, disconnect the battery and then switch on again. Note the effect on the lamp.

Fill in the missing words (5).

A relay consists of an .... which attracts an .... when .... Switch contacts in the relay .... and an external circuit is completed. The relay is ideal for using .... voltage circuits to operate .... voltage circuits.

▶ Relays also have contacts that are closed when the coil is carrying no current. These are called *normally closed contacts* (N_C).

42   INVESTIGATING ELECTRONICS

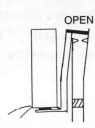
OPEN

▶ When the coil is energised, the contacts are pushed open by the action of the armature.

• •

E X E R C I S E S ▶ Build the circuit shown in the diagram.

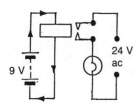

With the battery disconnected switch on the mains supply and note the effect on the lamp.

Switch off, connect the battery and switch on again. Note the effect on the lamp.

Fill in the missing words (6).

A normally closed contact will complete a circuit when the .... is .... When the .... is .... the contacts will .... and the circuit will be ....

• •

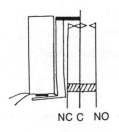
NC C NO

▶ Usually behind every set of open contacts is a set of closed contacts. The centre contact is called the common contact (C), and the one that it is touching when the coil is de-energised is the *normally closed contact* (N_C). The other contact is the *normally open contact* (N_O).

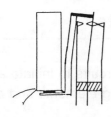

▶ When the coil is energised, the normally open contact closes and the normally closed contact opens. These are called *change-over contacts*.

DETECTOR CIRCUITS AND COMPONENTS

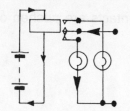

▶ This is another advantage of relays. They are able to control more than one circuit at the same instant in time. In this case one lamp switches off at the instant the other lamp switches on.

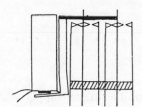

▶ Several change-over contacts can be stacked on to a relay until the required number of switching operations can be met.

EXERCISES ▶ Build the circuit shown in the diagram.

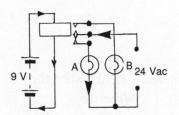

Switch on the circuit and note which lamp lights up.

Switch off the circuit and disconnect the battery. Switch on the circuit and note the effect on the lamps.

Fill in the missing words (7).

A change-over contact changes between a normally .... contact and a normally .... contact. The moving contact is called the .... contact. Change-over contacts can be used for a variety of .... applications at any .... in time.

Relays in detector circuits

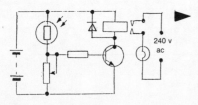

▶ A relay can be added to detector circuits to initiate alarm bells, motors, mains lighting, heating elements and other devices.

The diode is reverse biased across the relay coil to protect the transistor from damage due to the high *back e.m.f.* or reverse voltage created, as the magnetic field in the relay coil collapses when switched off.

EXERCISES

▶ Build the circuit shown in the diagram and test the operation of the circuit.

It is not very practical to light a lamp from an LDR. Modify the circuit to produce a useful circuit operated from an LDR.

Draw a diagram of the modified circuit and state its uses.

Build a circuit useful for lighting a porch light.

Fill in the missing words (8).

This basic light detector can become a dark detector by reversing the position of the .... and the .... or by using the .... closed .... on a .... The lamp can be replaced with ...., ... and other devices to produce a variety of useful circuits.

EXERCISES

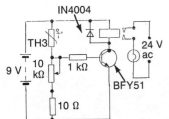

▶ Build the circuit shown in the diagram and test the operation of the circuit.

Lighting a mains lamp from a temperature source is not very useful. Modify the circuit to produce a more useful application.

Draw a diagram of the modified circuit and explain its uses.

Fill in the missing words (9).

The basic heat detector can become a .... detector by reversing the position of the .... and the .... or by using the .... closed .... on the ....

DETECTOR CIRCUITS AND COMPONENTS

▶ It is not always desirable for a relay to reset automatically. Take the case of an intruder who breaks a light beam to a photo-cell or switches on a light in a room protected by a light-detector alarm system. It is desirable for the alarm to ring until it is reset manually.

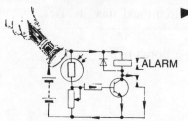

▶ If the intruder shines a light on the photo-cell the alarm rings as the transistor switches on and energises the relay. A second set of relay contacts called *retaining contacts*, close simultaneously and short circuit the collector/emitter of the transistor.

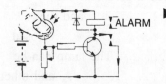

▶ Even though the light is now extinguished the relay remains energised through the second set of contacts, making the alarm independent of the detection part of the circuit.

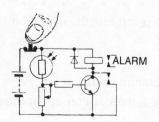

▶ A reset button has to be inserted into the relay supply circuit to cancel the alarm.

EXERCISES

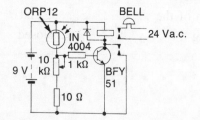

▶ Build the circuit shown in the diagram without the retaining contact.

Set the variable resistor so that the relay is off in natural light.

Connect the retaining circuit.

Shine a strong light on the LDR and note the effect on the alarm buzzer.

Remove the light and note the effect on the buzzer.

Connect a reset button into the circuit and cancel the bell.

Fill in the missing words (10).

The .... contacts bypass the ...., so that once .... the detection part of the circuit is .... of the alarm.

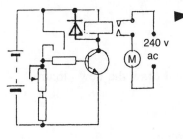

▶ A good practical circuit is used to detect the level of water in a tank and refill the tank when the water is low. The circuit shown in the diagram is a combination of a water-level detector and a relay circuit and does not suffice, as the pump motor will run when the tank is full.

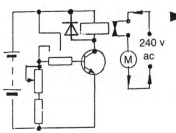

▶ Reversing the position of the variable resistor and the probes does not work in this case because no base current ever flows, but using normally closed contacts on the relay starts the motor when the tank is empty.

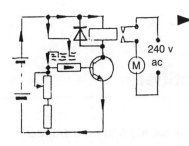

▶ In this circuit, the relay is energised all the time that the tank is full, which is wasteful of electricity over long periods and drastically reduces the life of the relay.

Two-stage detector circuits

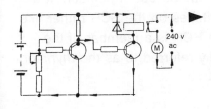

▶ A better idea is to invert the input information using a second transistor. If the probes are dry the first transistor does not switch on, allowing the base voltage of the second transistor to rise, switching it on and energising the relay.

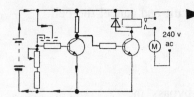

When the probes are submerged the first transistor conducts, shorting the base of the second transistor so that the relay does not energise. A two-stage circuit similar to this can also be used to detect heat and light by using the appropriate detectors.

E X E R C I S E S

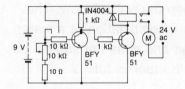

▶ Build the circuit shown in the diagram.

Test the operation of the circuit.

Adapt the circuit to another use and draw the new circuit diagram.

Explain the use of the adapted circuit.

Fill in the missing words (11).

In many cases a .... stage detector is superior to a .... stage detector. It is easier to set accurately and can be used to .... input information.

Silicon-controlled rectifiers

▶ The diagram shows the symbol for a *silicon-controlled rectifier* (SCR). The symbol resembles a normal rectifier diode with an *anode* and a *cathode* but in addition, it also has a *control gate* which switches on the SCR rather like the base of a transistor. The SCR is a member of the thyristor family, and is usually referred to as the *thyristor*.

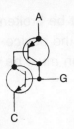
► The equivalent transistor circuit would look like the one shown. However, the top transistor is not the transistor that we have become familiar with. The arrow pointing towards the base tells us that this is a *PNP transistor* which is a reversal of the polarity of the more usual *NPN transistor*.

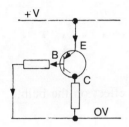
► The emitter is connected to the positive rail, so that the base can never be higher in potential than this. To switch on the transistor the base voltage must be lower than the emitter voltage so that current flows from the emitter to the base. This allows a larger emitter-to-collector current to flow and the collector voltage rises nearer to the supply voltage.

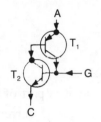
► If the gate is positive with respect to the emitter of T_2 then T_2 switches on, making its collector voltage drop. This lower collector voltage reduces the base voltage of T_1 making it conduct and increasing its collector voltage and hence the base voltage of T_2.

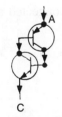
► The gate connection can now be removed and the SCR continues to conduct.

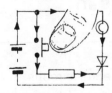
► Placed in a simple test circuit, the SCR may not appear to be more advantageous than a transistor but it has two advantages. Firstly they are extremely high-current devices allowing much larger load currents to be switched. Secondly once activated they latch (stay switched on), which can be useful in some applications.

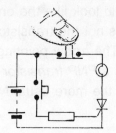

▶ To reset the circuit the anode line must be broken, this stops the current from flowing through the device and the latching effect is undone. The set button must be re-activated to re-latch the SCR.

EXERCISES

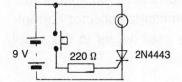

▶ Build the circuit shown in the diagram.

Press the start button and observe the effect on the bulb.

Release the start button and note the effect on the bulb.

Add a cancel button to the circuit and check its operation.

Fill in the missing words (12).

The SCR is a .... controlled .... more usually called the .... as it is part of the .... family. If the .... current is .... enough it will conduct, but once conduction takes place the .... can be disconnected and it will continue to ....

Silicon-controlled rectifiers in detector circuits

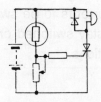

▶ A SCR can be used in any of the detector circuits previously dealt with. It is particularly useful for driving high-current loads, such as bells.

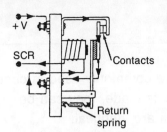

▶ The *bell* is basically a relay with a striker attached to the moving contact arm. Current flows through the coil, via the contacts, to energise it.

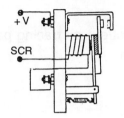

▶ When the coil is energised, the contacts open, breaking the coil circuit. The coil is now de-energised and returns to the contacts closed position ready for re-energising. Each time the contact arm falls back the striker hits the gong.

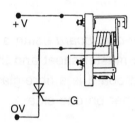

▶ A problem arises because, when the coil circuit is broken, the SCR unlatches and a single strike is all that is made.

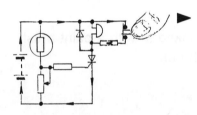

▶ To stop the SCR unlatching, a resistor is connected in parallel with the bell to supply a continuous current to the SCR. A cancel button can be introduced into this circuit to unlatch the SCR.

- -

EXERCISES ▶ Build the circuit shown in the diagram.

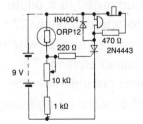

Use the variable resistor to set the circuit sensitivity. Expose the LDR to light and note the effect on the bell.

Operate the cancel button and note the effect on the bell.

Fill in the missing words (13).

A SCR can be used in any of the .... circuits. It is .... to small switch-on currents, but can handle large .... currents, making it useful for driving .... current devices.

- -

DETECTOR CIRCUITS AND COMPONENTS

Printed-circuit boards

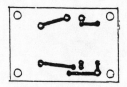

▶ *Printed-circuit boards* (PCBs) are commonly used in circuit manufacture and once a master is made, they can be manufactured in quantity and all are identical.

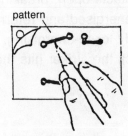

▶ The first step is to draw a master pattern on tracing paper showing the necessary track system.

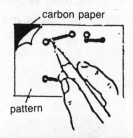

▶ The pattern is transferred to *copper-clad board* using a sheet of carbon paper between the master sheet and the copper-clad board. The copper-clad board is a fibre-glass board with a complete sheet of copper on one side instead of copper strips.

▶ The copper board pattern is inked in with a paint brush and etch-resist ink or an etch pen.

▶ When the ink is dry the board is submerged in a dilute ferric chloride mixture and heated to approximately 27 °C (80 °F). This solution dissolves the copper except where the ink protects it. This process is called *etching*. For proper etching to take place the ferric chloride solution must be kept in motion relative to the board by agitating the tray.

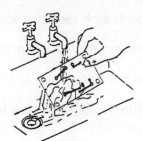

▶ The board is washed and dried.

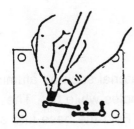

▶ The ink is removed by using a solvent cleaner or an ink eraser.

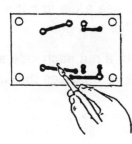

▶ Finally, a preservative flux will stop the copper from tarnishing and the holes can be drilled in the board, ready for the components to be mounted and soldered.

Practical exercise: continuity tester

PCB PRODUCTION

▶ Trace the PCB layout as shown in the diagram.

▶ Cut a piece of copper-clad board 45 mm × 28 mm.

▶ Transfer the layout pattern to the copper-clad board using carbon paper.

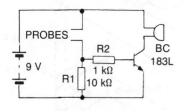

CIRCUIT DIAGRAM

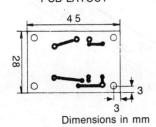

PCB LAYOUT

Dimensions in mm

DETECTOR CIRCUITS AND COMPONENTS

COMPONENT LAYOUT

- Fill in the areas to be retained with etch-resistant ink.
- Place in the etching bath to remove the unwanted copper.
- Clean off the ink and drill to suit the components' leads.
- Mount and solder the components paying attention to the correct polarities.

THE CASE AND ASSEMBLY

- Cut the base from a suitable material such as aluminium alloy or mild steel and drill the holes while flat. The 4.8 mm fixing holes can be left. Fold the base into shape.
- Using a piece of $\frac{1}{8}''$ welding rod, use a die to cut a 3 mm thread one end and grind the other to a point.
- Fasten the probe to the base with a nut each side and insulate the probe from the base as it passes through the hole. Fit a ring tag to the probe.
- Cut the lid and drill all the holes.

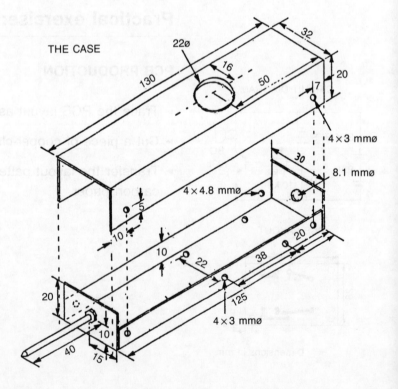

THE CASE

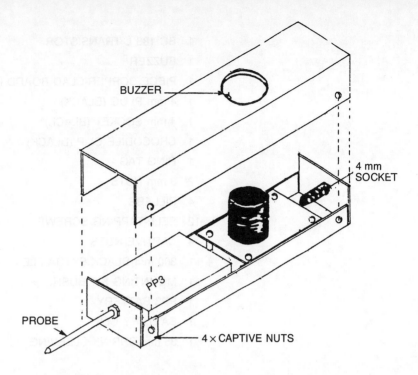

- Fold the lid and place over the base. Mark the 4.8 mm fixing holes through the lid-fixing holes and drill

- The lid can be painted if required.

- Fit the PCB into position using four 12.7 mm pillars.

- Fit a 4 mm socket to the back of the base.

- Manufacture a 0.5 m length of extra flex with a 4 mm plug on one end and a crocodile clip on the other.

- Connect the probe, battery, PCB, and socket as shown in the circuit diagram. Use a ring terminal for connecting to the probe.

- Add the No 4 captive nuts to the fixing holes and fasten the lid using No 4 self-tapping screws.

- A label can be manufactured and fitted to the lid if required.

COMPONENT LIST

1   1 kΩ RESISTOR
1   10 kΩ RESISTOR

DETECTOR CIRCUITS AND COMPONENTS

1 BC 183 L TRANSISTOR
1 BUZZER
1 PIECE COPPER-CLAD BOARD (45 mm × 25 mm)
1 4 mm PLUG (BLACK)
1 4 mm SOCKET (BLACK)
1 CROCODILE CLIP (BLACK)
1 RING TAG
2 3 mm NUTS
4 PILLARS
12 SELF-TAPPING SCREWS
4 CAPTIVE NUTS
1 300 mm BLACK EXTRA FLEX
1 MOUNTING KIT BUSH
1 PP3 BATTERY
1 BATTERY CLIP
1 35 mm SHRINK SLEEVING

Self-assessment questions

1. Voltage is calculated from the formula

    (a) $V = I + R$
    (b) $V = R + I$
    (c) $V = IR$
    (d) $V = I^2R$

2. If two ammeters are connected in series A1 will

    (a) Read the same as A2
    (b) Read more than A2
    (c) Read less than A2
    (d) Could read more or less

3. In a voltage divider V1 will

    (a) Read the same as V2
    (b) Read more than V2
    (c) Read less than V2
    (d) Could read more or less

4. An LDR has a high resistance in

   (a) Good light
   (b) Poor light
   (c) High temperatures
   (d) Low temperatures

5. A thermistor has a low resistance in

   (a) Good light
   (b) Poor light
   (c) High temperatures
   (d) Low temperatures

6. A relay is a

   (a) Permanent magnet device
   (b) Electro mechanical device
   (c) A solid state device
   (d) A moving coil device

7. A relay is used to

   (a) Interface high voltage circuits to low voltage circuits
   (b) Perform a number of tasks at the same moment
   (c) Both A and B
   (d) Neither A or B

8. A reverse biased diode is connected across a coil to

   (a) Rectify the coil voltage
   (b) Bypass the coil
   (c) Block the coil current
   (d) Dissipate the back e.m.f. from the coil

9. An SCR is a

   (a) Transistor
   (b) Rectifier diode
   (c) Thyristor
   (d) Triac

10. The advantage of an SCR is

   (a) It latches when switched on
   (b) It is a high current device
   (c) Both A and B
   (d) Neither A or B

Answers to self-assessment questions

1, (c);   2, (a);   3, (d);   4, (b);   5, (c);   6, (b);   7, (c);
8, (d);   9, (c);   10, (c).

Missing words

(1) Variable, sensitivity, high.
(2) Sensitive, transistor, transistor, variable resistor, base, transistor.
(3) Hot, frost, cookers, refrigerators.
(4) Fluid, fluid level detector, detect, fluid, continuity.
(5) Electromagnet, armature, energised, close, low, higher.
(6) Coil, de-energised, coil, energised, open, broken.
(7) Open, closed, common, switching, moment.
(8) LDR, variable resistor, normally, contact, relay, bells, motors.
(9) Cold, thermistor, variable resistor, normally, contact, relay.
(10) Retaining, transistor, energised, independent.
(11) Two, single, invert.
(12) Silicon, rectifier, thyristor, thyristor, gate, high, gate, conduct.
(13) Detector, sensitive, load, high.

CHAPTER 3

AMPLIFICATION

Voltage amplifiers

The purpose of a voltage amplifier is to produce a signal at the output which is an exact copy of the signal at the input but at a higher voltage level.

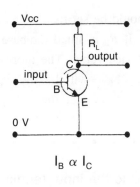

$I_B \propto I_C$

$I_C \propto 1/V_{CE}$

▶ In the case of a *voltage amplifier*, the base of the transistor is the input and the collector is the output. Before applying a signal to the base, the collector voltage is set to $V_{CC}/2$ volts (e.g. if V_{CC} = 9 V then the collector voltage must equal 9/2 = 4.5 V).

▶ The base current I_B is directly proportional to the collector current I_C. When I_B rises then I_C also rises.

▶ When I_C rises, the voltage drop across R_L, (V_{RL}), rises and since $V_{CC} = V_{RL} + V_{CE}$, V_{CE} falls. This results in the output voltage, V_{CE}, being *inversely proportional* to I_C.

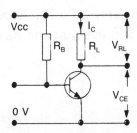

▶ The base current I_B is dependent on the resistance value of R_B; if R_B is a high value I_B will be low and vice versa. It is possible to set the output voltage V_{CE} to $V_{CC}/2$ by adjusting the value of R_B. This is called *biasing*.

The value of the load resistor R_L is usually in the region of 1 kΩ to 4.5 kΩ and is used to form a voltage divider with the collector/emitter resistance of the transistor.

d.c. gain

$$h_{FE} = I_C/I_B$$

▶ The ratio of I_C to I_B is called the *forward current transfer ratio*, h_{FE} and is a measure of the *gain* of a transistor. Thus when I_C is twice I_B, the gain is 2.

- -

EXERCISES

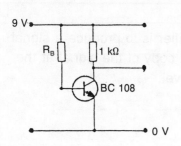

▶ Choose a value for R_B which will bias V_{CE} to 2.5 V d.c. (Note R_B must be at least 10 times greater than R_L to avoid transistor damage.)

Build the circuit shown in the diagram.

Measure the base current, base voltage and output voltage.

Fill in the missing words (1).

The .... voltage V_{CE} should be set to .... the supply voltage. If R_B is small the base current will be .... and vice versa. The more base current flowing, the .... will be the output voltage, so the value of R_B will .... the amplifier stage.

- -

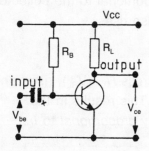

▶ When an a.c. signal, V_{be}, is applied to the input terminals, an amplified voltage, V_{ce}, is obtained at the output terminals.

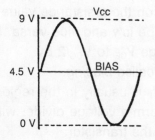

▶ The input waveform need not necessarily be a sine wave, any wave shape may be amplified and every sound has its own unique waveshape.

60   INVESTIGATING ELECTRONICS

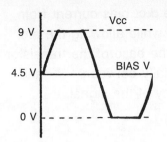

▶ When the input signal is too large for the amplifier stage to handle, the output signal becomes distorted. The transistor is then said to be *overdriven*. The output signal is then *not* an exact copy of the input signal and *amplitude distortion* is said to occur.

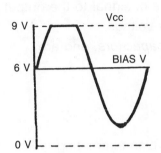

▶ Amplitude distortion can also be caused by incorrect biasing. With a supply of 9 V and 6 V biasing at the output, the amplified waveform is flattened at the top when the amplitude of the amplified waveform exceeds 3 V.

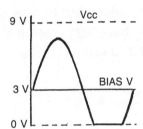

▶ With a 3 V bias at the output and a supply of 9 V, amplitude distortion occurs when the amplified waveform exceeds 3 V, due to the waveform flattening at the bottom.

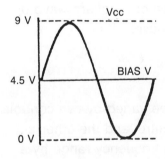

▶ When biasing is correct, then it is possible for the output signal to be undistorted up to a value of 4.5 V giving 9 V peak to peak.

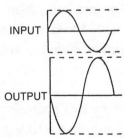

▶ In addition to amplifying the a.c. input signal, the amplifier stage is also an inverter. A single stage amplifier produces an output waveform which is 180° out of phase with its input waveform.

AMPLIFICATION   61

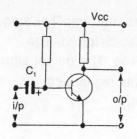

▶ Capacitor C_1 is used to block the d.c. bias current from flowing back into the input stage while allowing the a.c. signal to pass from the input to the base of the transistor. Its value is not critical and is usually between 1 μF and 10 μF depending on the frequency of the signal.

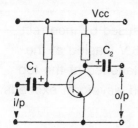

▶ Capacitor C_2 is used to pass the a.c. signal to the output stage while blocking the d.c. bias voltage.

C_1 and C_2 are called *coupling capacitors*, and are usually electrolytic capacitors.

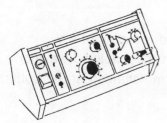

▶ A *signal generator* is used to produce alternating waveforms of varying shapes, amplitudes, and frequencies for testing amplifiers. The range of these instruments is normally from a few hertz to about 100 kHz.

▶ A signal select switch is used to choose the type of signal required. This may be a sine wave or a square wave or maybe a triangular or saw-tooth wave.

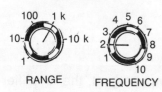

▶ The *frequency* of the signal can be varied by two controls. One is a fine frequency speed control and the other a multiplier control which alters the frequency range by a factor of 10 each time it is turned. For example, 2 on the fine control in the diagram together with 1 kHz selected on the range switch results in a frequency of about 2 kHz.

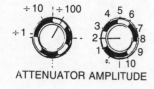

▶ The *amplitude* of the waveform is set by two controls, an amplitude control which gives voltages from 0–10 V and an *attenuator* control which divides the amplitude setting by 1, 10 or 100. From the settings in the diagram the amplitude of the waveform is 2 ÷ 100 = 0.02 V.

Oscilloscopes

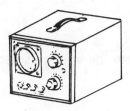

▶ In order to observe alternating waveforms the *oscilloscope* is used. It displays the amplitude and frequency of a waveform over a set period of time.

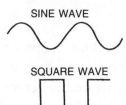

SINE WAVE

SQUARE WAVE

▶ A waveform can take on a variety of shapes but all have one thing in common, with the exception of direct current, they all vary in magnitude and frequency.

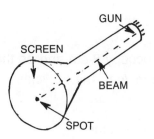

▶ The oscilloscope uses a *cathode ray tube* to display the waveform. A stream of electrons is fired from an electron gun at the back of the tube to form a spot on the screen at the front of the tube.

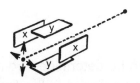

▶ The spot can be deflected horizontally or vertically by electrically biasing two sets of plates inside the tube neck. The Y plates send the spot up or down and the X plates send the spot to the left or to the right.

INTENSITY

MIN   MAX

▶ If the spot is left stationary and at maximum intensity it can damage the screen lining. So before switching on, the brightness control should be set to the minimum.

X SHIFT   Y SHIFT

MIN   MAX MIN   MAX

▶ When the X or Y shift controls are set near to their minimum or their maximum settings the trace will probably disappear off one of the edges of the screen. To ensure that the trace appears on the screen the shift controls should be set half way.

AMPLIFICATION   63

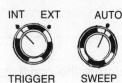

▶ The timebase is triggered internally so the trigger select switch is set to internal and the sweep control to auto.

▶ If we intend to test an a.c. waveform the voltage select switch must be set to a.c.

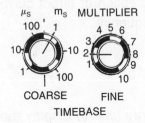

▶ The oscilloscope has two *timebase* controls, namely a coarse control and a fine control. The coarse control is marked in time per centimetre divisions on the screen and the fine control is a multiplier of the first. Thus the setting shown is 1 ms × 2 = 2 ms/cm.

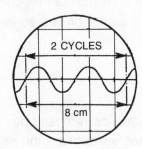

▶ The 2 cycles of waveform shown occupy 8 cm and as the timebase control settings are at 2 ms/cm, the 2 cycles represent:

2 ms/cm × 8 cm = 16 ms of time

FREQUENCY = $\frac{\text{No OF CYCLES}}{\text{TIME}}$

▶ The frequency of the waveform can be determined from the relationship shown. The unit of frequency is the *hertz* (Hz). Thus, the frequency of the waveform is:

$2/(16 \times 10^{-3})$ = 125 Hz

▶ The oscilloscope also has two voltage gain controls, namely a coarse control and a fine control. The coarse control is marked in volts per centimetre. In this case each vertical centimetre on the screen will display one volt of the waveform. Multiply this by the number of centimetres that the waveform rises, to calculate the peak voltage of the waveform.

INVESTIGATING ELECTRONICS

▶ The fine control is used to calibrate the display to match the voltage of a known input signal, or simply to increase the height of the waveform to a convenient level on the screen.

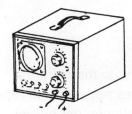

▶ Circuits to be tested are connected to the input terminals and the oscilloscope is switched on. The on/off switch is quite often incorporated into the brightness control.

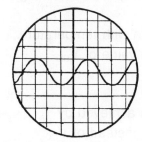

▶ When the oscilloscope has warmed up turn up the brightness control until the trace appears and centralise it using the X and Y shift controls.

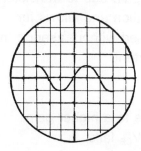

▶ If the trace is too short increase the X gain control until the trace travels the whole width of the screen.

The trace can be trimmed to a fine line by adjusting the focus control.

EXERCISES

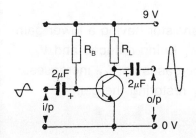

▶ Build the circuit shown. Connect a signal generator to the input. Set the signal generator to give a sinewave of amplitude 0.01 V. Connect an oscilloscope across the output. Is the output a sine wave?

Measure the amplitude of the output waveform using the oscilloscope. Calculate the gain.

Using a twin-beam oscilloscope, check the input and output waveforms simultaneously. Determine the phase relationship of the two waveforms.

AMPLIFICATION

Fill in the missing words (2).

An alternating .... can be ....
when passed through an .... stage. The
output signal is .... degrees ....
with the input signal. Capacitors are used to separate the
.... from the ....

- -

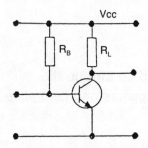

▶ The type of biasing shown is tailored to the gain of a particular transistor and if it becomes necessary to replace this transistor with one of a different gain, the biasing voltage of V_{CE} will be affected.

Transistors also change their gain with temperature change which again alters the bias value of V_{CE}.

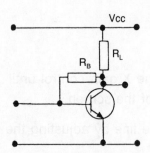

▶ A much better method of biasing is achieved by connecting the base resistor to V_{CE} instead of V_{CC}.

This compensates for increased gain due to transistors changing temperature during operation and allows for replacement of transistors being used without altering the biasing voltage V_{CE} to any great extent.

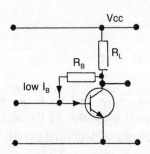

▶ As the collector current I_C rises due to a higher gain replacement transistor or heating, V_{CE} falls.

When V_{CE} falls, the base current I_B decreases and as I_C is equal to $I_B \times h_{FE}$, I_C decreases allowing V_{CE} to rise again.

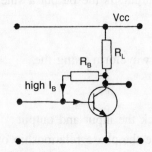

▶ Alternatively, a replacement transistor having a lower gain than its predecessor results in I_C being reduced and V_{CE} being increased. However I_B increases as V_{CE} increases, causing I_C to increase, partially restoring the gain.

INVESTIGATING ELECTRONICS

EXERCISES

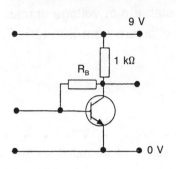

▶ Build the circuit shown in the diagram. Choose a value for R_B to give a 4.5 V output.

Change the transistor and measure the output voltage.

Fill in the missing words (3).

When the base resistor is connected to the collector instead of the ...., changes in .... due to the transistor replacement and .... are kept ....

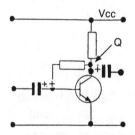

▶ When R_B is connected to the collector of a transistor, it is sometimes referred to as a *feedback biased* amplifier.

The feedback biased amplifier is quite a good practical amplifier but some variation of V_{CE} still exists and can become a problem when the amplified signal is large.

The d.c. voltage between the collector and 0 V, (i.e. V_{CE}), is usually referred to as the *quiescent* or *Q point*.

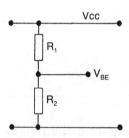

▶ An improvement on the feedback amplifier is the *fully stabilised amplifier* which will stabilise the Q point irrespective of gain or temperature.

Instead of the single resistor R_B previously used to set the base current, a potential divider, comprising R_1 and R_2, is used to set the base voltage V_{BE}.

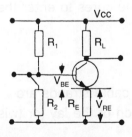

▶ The base/emitter voltage of a transistor is usually between 0.6 and 0.7 V. An emitter resistor R_E is used to bring the voltage $V_{BE} + V_{RE}$ up to about 1.6 V. The value of R_E may be calculated using $R_E = V_E \div I_C$. So when I_C is, say 1 mA, (0.001 A), and V_E is 1 V, then $R_E = 1 \div 0.001 = 1\ \text{k}\Omega$.

AMPLIFICATION

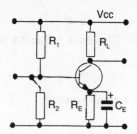

▶ The input signal across R_2 also affects the voltage across R_E and since this is undesirable, capacitor C_E, which passes a.c., is used to keep a stable d.c. voltage across R_E. Typical values for this capacitor are from 20–100 μF.

EXERCISES

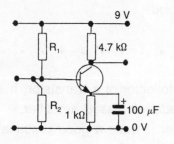

▶ Build the circuit shown, choosing values of R_1 and R_2 to give a Q point of 4.5 V. (Remember that R_1 must be at least $10 \times R_L$ to avoid transistor damage.)

Replace the transistor with another and note the effect on the Q point voltage.

Fill in the missing words (4).

The transistor bias of a fully stabilised amplifier is set by the resistors ...., ... and .... C_E is a bypass capacitor to separate the .... and the .... at the emitter.

Terminal blocks

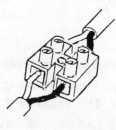

▶ Very often it becomes necessary to joint or terminate cables using a *terminal block*. The cable sheath is removed just sufficiently to allow the cores to enter the terminals.

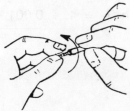

▶ After stripping the cable ends, the cable strands are twisted tightly in the natural direction of the lay (or twist) of the cable.

▶ When twisted tightly the cores should be trimmed to slightly less than half of the width of the terminal block.

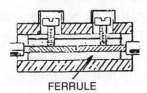

FERRULE

▶ The cores should touch in the middle of the terminal block and the insulation should enter the insulated part of the terminal block but not the metal ferrule inside.

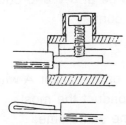

▶ Single-core cable can be weakened when a screw is tightened on it when terminating.

For this reason the cable is doubled back before terminating and this reduces the damage.

TERMINAL BLOCK AMPLIFIER

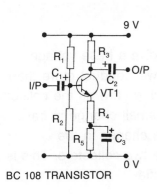

BC 108 TRANSISTOR

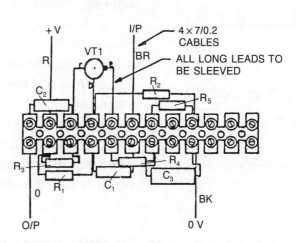

- -

E X E R C I S E S ▶ Build the circuit shown using a terminal block, with:

$R_1 = 18$ kΩ; $R_2 = 18$ kΩ; $R_3 = 47$ kΩ;
$R_4 = 470$ Ω; $C_1 = 1$ μF; $c_2 = 2$ μF;
$C_3 = 22$ μF; $VT_1 =$ a BC 108 transistor.

Sleeve all exposed component leads.

Test the circuit using a twin-beam oscilloscope.

- -

Analogue microelectronics

OPERATIONAL AMPLIFIERS

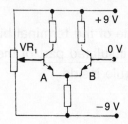

▶ The circuit in the diagram is a *differential amplifier* sometimes called a *long-tailed pair*. It uses both positive and negative power supply lines. The base of transistor B is tied down to 0 V, the mid-voltage between the two supply lines. When VR_1 is set exactly half way then the base voltage of transistor A will also be at 0 V and both transistors will conduct an equal amount.

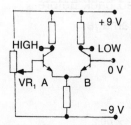

▶ With VR_1 set to give a base voltage on transistor A which is lower than 0 V, transistor B will conduct the most, making its collector output low. At the same time, transistor A will conduct less so that its collector output will rise.

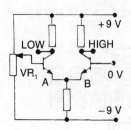

▶ It follows that when VR_1 is set to give a base voltage higher than 0 V then transistor A will conduct more than transistor B so that its collector will go low as the collector of transistor B goes high. Thus a small change in base voltage results in a relatively large change in the difference in voltages between the two collectors. This is why it is called a differential amplifier.

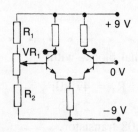

▶ Transistor A could be damaged by reducing its base resistance to a very low value, so R_1 and R_2 are added to limit the base current. If they are the same value they will have no effect on the biasing of transistor A.

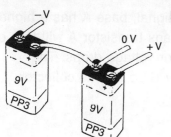

▶ Differential amplifiers require both positive and negative power supplies. It is possible to create a positive and negative power supply from two standard power supplies or two batteries by linking the positive of one supply to the negative of the next, as shown for two batteries.

EXERCISES

▶ Build the circuit shown in the diagram.

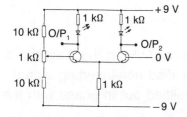

Move the potentiometer wiper towards +V and note the effect on the LEDs.

Measure the two voltages at O/P_1 and O/P_2.

Move the potentiometer wiper towards −V and note the affect on the LEDs.

Measure the two voltages at O/P_1 and O/P_2.

Fill in the missing words (5).

A differential amplifier amplifies the .... between two base inputs. This results in a relatively .... change in the difference in voltages at the .... of the two transistors.

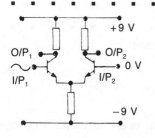

▶ When an alternating signal is applied to I/P_1 both the transistors will switch on and off in sympathy with the signal.

AMPLIFICATION 71

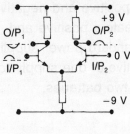

On the rising half-cycle of the signal, base A has a higher voltage than base B, which means transistor A will turn on more than B, so that its collector voltage drops. The reverse will happen to transistor B so that its collector voltage rises.

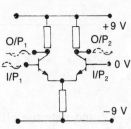

For the second half cycle, I/P$_1$ voltage is lower than I/P$_2$, resulting in O/P$_1$ voltage rising and O/P$_2$ voltage dropping. O/P$_2$ is an amplified version of I/P$_1$, whereas O/P$_1$ is an amplified *and inverted* version of I/P$_1$.

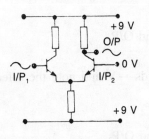

When only O/P$_2$ is used, shown as O/P on the diagram, a signal on I/P$_1$ results in an amplified non-inverting output, that is, the output signal is amplified but in phase with the input signal.

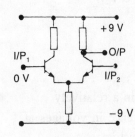

If input 1 is connected to 0 V and the signal fed into input 2 then the output will be inverted, input 2 then is the inverting input.

EXERCISES

Build the circuit shown in the diagram.

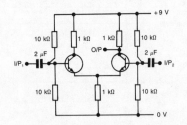

Connect I/P$_2$ to zero volts.

Apply a small sine wave to I/P$_1$ and note the output in phase and amplitude.

Connect I/P$_1$ to zero volts.

Apply the same sine wave to I/P$_2$ and note the output in phase and amplitude.

Determine the value of h_{FE} for the amplifier.

Fill in the missing words (6).

The .... amplifier can be connected as a .... amplifier or an .... amplifier. When I/P$_1$ is connected to .... V, the amplifier is an .... amplifier, but if I/P$_2$ is connected to .... V, then it becomes a .... amplifier.

● ●

The long-tailed pair is a relatively simple form of differential amplifier, but more complex circuits are used, as can be seen in the diagram, these being called *integrated circuit operational amplifiers*.

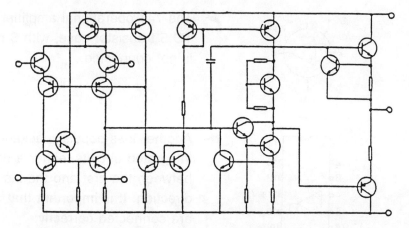

Integrated circuit operational amplifiers are now so convenient and cheap that it is hardly worth while to build a discrete component amplifier.

The diagram shows the discrete equivalent of the type 741 operational amplifier which is probably the cheapest and most widely used of the operational amplifiers. The operational amplifier should be treated as a building block and the internal components may be ignored.

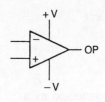

▶ The symbol used to denote an operational amplifier is as shown. It has two inputs, an inverting input (−) and a non-inverting input (+). When the non-inverting input is made more positive then the output becomes more positive, but when the inverting input is made more positive, the output becomes more negative.

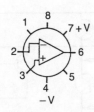

▶ The most common type of 741 package is the eight-pin *dual in-line* package (DIL). The circuit symbol is shown drawn on the package. The positive and negative supply connections are rarely shown on a circuit diagram, but they must be correctly connected or the operational amplifiers will not operate.

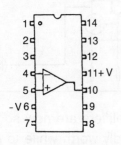

▶ The 741 operational amplifier is also manufactured in a TO 5 transistor case, with 8 pins. Pin 8 on both 741 types is not connected.

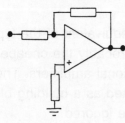

▶ Another less popular package is the 14-pin DIL. On all DIL integrated circuits, either a dot marks pin 1 or a notch between the first and the last pins denotes the pinning direction. It is important that all integrated circuits (ICs), are connected correctly.

▶ The + and − signs shown in the diagram do not refer to the power-supply lines but to the inverting (−) and non-inverting (+) inputs. The signal is connected to the inverting input making this configuration an *inverting amplifier*.

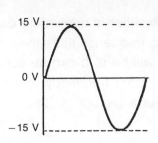

▶ Because this amplifier utilises positive and negative supplies the signal does not vary about a Q point, as in the discrete amplifiers but varies about 0 V, the half way voltage between the two supply rails. The 741 operational amplifiers may be used on power supplies between ±3 V to ±18 V, but it is generally thought safer not to exceed ±15 V for fear of risking damage to the amplifier.

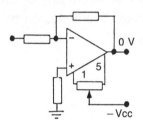

▶ The output can be set to zero by varying a potentiometer connected between pins 1 and 5. This action is called the *offset null*, which is a means of zeroing any positive or negative voltages that may appear at the output before a signal is applied.

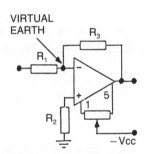

▶ The gain of an operational amplifier is extremely high and to reduce the gain, external resistors are needed. The high gain signal at the output is fed back to the input via the resistor R_3 almost cancelling the input signal and for this reason this point is called the *virtual earth*.

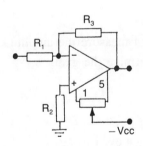

▶ The gain is now dependent on the ratio of R_3 to R_1, so that if R_3 is ten times greater than R_1 then the operational amplifier has a gain of ten. R_2 should be equal to the equivalent parallel resistance of R_1 and R_3.

Thus for $R_1 = 1 \text{ k}\Omega$ and $R_3 = 10 \text{ k}\Omega$,

$$R_2 = \frac{1 \times 10}{1 + 10} \simeq 909 \text{ }\Omega.$$

AMPLIFICATION

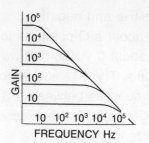

▶ The gain of the 741 operational amplifier is dependent on the frequency of the signal. As the frequency rises the gain falls until at 1 MHz, the gain will be 0. It can be seen that lower gains are stable over a larger frequency range or bandwidth than higher gains. For a gain of 10, the bandwidth is about 10^5 Hz.

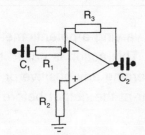

▶ Coupling capacitors can be used to separate any offset voltages from the signal, which makes the inclusion of the offset null potentiometer unnecessary.

E X E R C I S E S

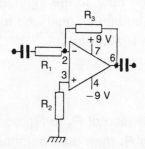

▶ Calculate the values of R_1, R_2 and R_3 needed to give an overall gain of 180 at optimum frequency.

Build the circuit shown in the diagram.

Add a 10 kΩ offset null control and zero the output voltage at pin 6.

Amplify a 1 kHz, 1 mV sine wave and measure the amplitude of the output on an oscilloscope. Determine the gain.

Fill in the missing words (7).

Very high .... can be achieved using .... amplifiers. The .... is set by the relationship between resistors .... and ....

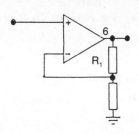

▶ In the *non-inverting amplifier* configuration shown, the signal is applied to the non-inverting input. A rise in the input voltage will cause an amplified rise at the output pin 6. To set the gain, part of the output is fed back to the inverting input via resistor R_1.

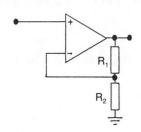

▶ If R_1 is kept much larger than R_2, the gain will be approximately equal to $R_1 \div R_2$. R_1 can be increased for higher gains and may be varied between 100 kΩ and 1 MΩ. Alternatively, R_2 can be decreased for higher gains and may be varied between 10 kΩ and 100 kΩ.

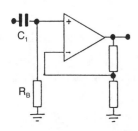

▶ When the input signal is purely a.c. and a coupling capacitor C_1 is used, then an input bias resistor is necessary. R_B should be as high as possible but its value is limited by the amount of offset voltage at the output.

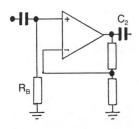

▶ A second coupling capacitor C_2 stops the offset voltage being transmitted to the next stage, enabling a much higher value of R_B to be used. In this case, its value can be increased by a factor of 10–20. Without C_2, R_B should be limited to 100 kΩ or less, dependent on gain.

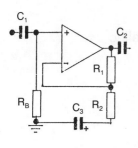

▶ Another good technique for reducing the offset voltage is to add a capacitor C_3 in series with R_2, then R_B can be made as high as R_1 without introducing any serious offset voltage.
   Typical values of C_3 are between 1–100 μF.

AMPLIFICATION   77

EXERCISES ▶ Build the circuit shown in the diagram.

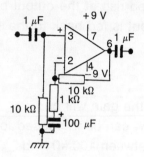

Apply a 1 kHz, 10 mV signal to the input and using an oscilloscope, measure the output voltage.

Measure the offset voltage at the output (pin 6) with no signal input.

Remove the 100 μF capacitor and note the change in offset voltage at the output.

Fill in the missing words (8).

The offset .... of an .... can be reduced by adding a .... in series with the .... resistors.

Self-assessment questions

1. The d.c. output voltage of an amplifier is set to

    (a) V_{CC}
    (b) $V_{CC} \div 2$
    (c) $V_{CC} \times 2$
    (d) $V_{CC} \div 4$

2. Setting the output voltage is called

    (a) Baseing
    (b) Biasing
    (c) Voltage dividing
    (d) Amplifying

3. Coupling capacitors are used to

    (a) Block the a.c. waveform and the d.c. bias
    (b) Block the a.c. waveform and pass the d.c. bias
    (c) Block the d.c. bias and pass the a.c. waveform
    (d) Pass both the a.c. waveform and the d.c. bias

4. Negative feedback

(a) Causes oscillation
(b) Increases gain
(c) Reduces gain
(d) Stops oscillation

5. A bypass capacitor

(a) Keeps the emitter voltage constant
(b) Passes d.c. to earth
(c) Couples one stage to another
(d) Blocks the a.c. from earth

6. A long-tail pair is a

(a) Oscillator
(b) Small-signal amplifier
(c) Operational amplifier
(d) Differential amplifier

7. A waveform flattened at the top only is

(a) Overdriven
(b) Biased too high
(c) Biased too low
(d) Correctly biased

8. A waveform flattened at the top and bottom is

(a) Overdriven
(b) Biased too high
(c) Biased too low
(d) Correctly biased

9. The offset null control

(a) Reduces the gain to zero
(b) Sets the output to zero volts
(c) Sets the input to zero volts
(d) Nullifies the waveform

10. Operational amplifiers can be connected as

(a) Inverting but not non-inverting amplifiers
(b) Non-inverting but not inverting amplifiers
(c) Neither inverting nor non-inverting amplifiers
(d) Both inverting and non-inverting amplifiers

Answers to self-assessment questions

1. (b); 2. (b); 3. (c); 4. (c); 5. (a); 6. (d); 7. (b);
8. (a); 9. (b); 10. (d).

Missing words

(1) Output, half, high, lower, biases.
(2) Waveform, amplified, amplifier, 180, out of phase, d.c., a.c.
(3) Supply, gain, temperature, small.
(4) R_1, R_2, R_E, a.c., d.c.
(5) Difference, large, collectors.
(6) Differential, non-inverting, inverting, 0, inverting, 0, non-inverting.
(7) Gain, operational, gain, R_3, R_1.
(8) Voltage, operational amplifier, capacitor, feedback.

CHAPTER 4

POWER SUPPLY UNITS

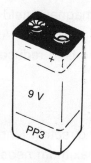

▶ The circuits that have been built so far have been powered by *batteries* which are ideal for portable equipment.

▶ However, batteries deteriorate and become inoperative as the chemicals used to produce a voltage become exhausted.

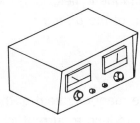

▶ When the equipment to be powered is permanently within a domestic or industrial environment, then a *power supply unit* can be used to replace the battery.

Transformers

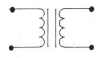

▶ One of the principal components in a power supply unit (PSU), is a *transformer*. This is used to reduce the 240 V, a.c. supply to the lower voltage required for a PSU.

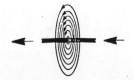

▶ When a current is flowing in a conductor a *magnetic field* is generated which encircles the conductor, as shown. The size of this magnetic field is determined by the amount of current flowing in the conductor.

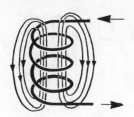
▶ Winding the conductor to form a *coil* causes the magnetic fields produced around each turn to amalgamate to produce one large magnetic field around the coil. The magnetic field can be further strengthened by inserting an iron core in the centre of the coil.

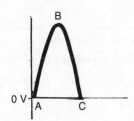
▶ When the coil is connected to an a.c. supply, as the current increases the strength of the magnetic field also increases until a maximum is reached at point B. As the current falls back to zero, the magnetic field reduces to zero.

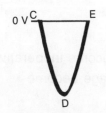
▶ The current now flows in the opposite direction (shown as a negative value) until it reaches its maximum negative level at point D, and then rises back to zero. This can be represented by a sine wave.

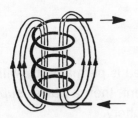
▶ The magnetic field again follows the amount of the current flowing, increasing to its maximum strength as the current reaches point D, but this time the magnetic field is in the opposite direction. This is the action of the *primary winding* of a transformer.

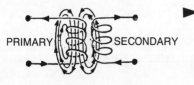
▶ If a second coil is passed inside the magnetic field of the primary coil the magnetic field will produce a current in the *secondary winding*.

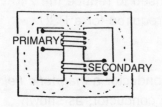
▶ The magnetic field is reinforced and made stronger when it is passing through iron, so both the coils are wound on a laminated iron core, as shown.

82   INVESTIGATING ELECTRONICS

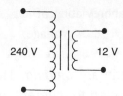

▶ A transformer has a proportional volts-to-turns ratio. Therefore, when there are fewer turns on the secondary winding than the primary, then the secondary voltage will be smaller than the primary voltage, making a *step-down transformer*.

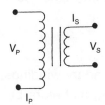

▶ There is an inverse relationship between voltage and current and so if the secondary voltage (V_S) is reduced the secondary current (I_S) is increased proportionally. This is because the primary power is equal to the secondary power and for alternating supplies, power is proportional to both voltage and current.

$$V_P \times I_P = V_S \times I_S$$

▶ The relationship between primary and secondary currents and voltages is as shown. By transposing this formula when three of the four quantities are known, it is possible to calculate the fourth.

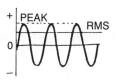

▶ The output voltage of a transformer read on an a.c. voltmeter is the root mean square voltage or *r.m.s. voltage*, this being the effective voltage of an a.c. sine wave, taking into consideration the increase and decrease of the a.c. voltage. The r.m.s. value of a sine wave is 0.707 of the *peak value*.

- -

E X E R C I S E S ▶ Build the circuit shown in the diagram.

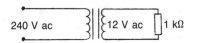

Measure the transformer output with an a.c. voltmeter.

Calculate the peak voltages of both the primary and secondary windings.

Fill in the missing words (1).

An a.c. .... reads the .... voltage but an oscilloscope shows a .... to

POWER SUPPLY UNITS

.... value. r.m.s. is the abbreviation of
.... and is the .... voltage.

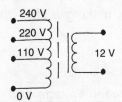

▶ Some transformers have a *tapped* primary *winding*. This enables the transformer to be connected to different mains supplies. In this country the supply varies from area to area between 220 and 240 volts, in some other countries the supply may be 110 volts.

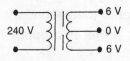

▶ Some transformers have tapped secondary windings, these can be used as a single winding as before by neglecting the centre tap, (12 V), or as two separate 6 V windings. At any moment in time one winding will be positive with respect to the centre tap and the other negative.

EXERCISES ▶ Build the circuit shown in the diagram.

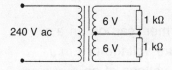

Measure each output using an a.c. voltmeter.

Measure the output voltage across the windings when connected in series.

Fill in the missing words (2).

A .... transformer provides two outputs, one .... with respect to the .... and one .... with respect to the .... The two windings can be connected in series to give .... the output voltage.

Rectifiers

▶ The low voltage a.c. output from the transformer needs to be changed to a direct current supply. This is called *rectification*. The simplest rectifier circuit is the *diode*.

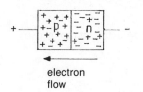

electron flow

▶ The diode has a *p–n junction*. When the p region is connected to the positive terminal of the supply and the n region to the negative of the supply, the negatively charged electrons in the n region are drawn through the p region to the positive terminal, making the diode behave as a conductor.

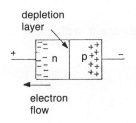

depletion layer

electron flow

▶ If now the diode is re-connected with the p region to the negative terminal of the supply and the n region to the positive of the supply, the electrons in the n region are drawn away from the diode junction, causing a depletion layer and making the diode act as an insulator.

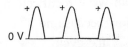

▶ As the a.c. output from the transformer provides a + voltage to the p region during one half cycle and a − voltage on the second half cycle, only half of the cycle will be allowed to pass through the diode and the other half will be blocked, resulting in the waveform shown. This is called *half-wave rectification*.

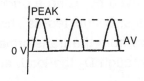

▶ The effective voltage will not be the peak voltage but the *average voltage*. Thus for a sine wave;

average voltage = (peak voltage × 0.637) ÷ 2

For a peak voltage of 37.7 V, the average voltage is

(37.7 × 0.637) ÷ 2 = 12 V

A d.c. voltmeter indicates the average value.

POWER SUPPLY UNITS

EXERCISES ▶ Build the circuit shown in the diagram.

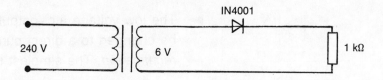

Measure the output voltage of the transformer using an a.c. voltmeter.

Measure the voltage across the 1 kΩ resistor using a d.c. voltmeter.

Observe the load voltage using the oscilloscope, and measure the peak voltage. Sketch the waveform.

Calculate the average voltage and compare it with the d.c. voltmeter reading.

Fill in the missing words (3).

A d.c. voltmeter reads the .... voltage which is the .... voltage. On half-wave rectification the .... voltage will be low because no current flows at all for .... the cycle.

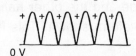

▶ A better waveform for use in a PSU is obtained by using a full-wave rectifier.

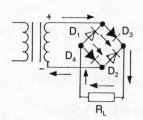

▶ *Full-wave rectification* can be achieved by using a *bridge rectifier*. This consists of four diodes, connected as shown. On the first half cycle current flows from the transformer + in the direction of the arrows through D_3 the load, and D_4, to the transformer −.

86 INVESTIGATING ELECTRONICS

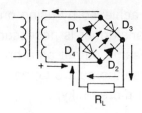

▶ On the second half cycle, current flows from the transformer + which is now at the bottom of the winding, through D_2, the load and D_1 and back to the transformer −. Because the current of each half cycle flows through the load resistor in the same direction the output is always positive.

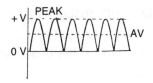

▶ The effective d.c. voltage at the output of a bridge rectifier as read on a d.c. voltmeter is the average value and is 0.637 of the peak value.

The peak value needed to give a 12 V effective value is thus reduced from 37.7 V with a half-wave rectifier to 18.85 V with a full-wave rectifier.

■ ■

EXERCISES ▶ Build the circuit shown in the diagram.

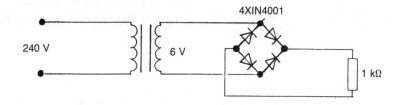

Measure the load voltage using a d.c. voltmeter.

Measure the peak load voltage using an oscilloscope.

Calculate the average d.c. voltage and compare it with the d.c. voltmeter reading.

Fill in the missing words (4).

The effect of full-wave rectification compared with half-wave rectification is to .... the voltage required at the .... output, to give the same .... voltage. Full-wave bridge rectification needs .... diodes.

■ ■

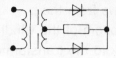
▶ Full-wave rectification can also be achieved by using only two diodes, but a centre-tapped transformer is necessary. Although this circuit produces the same output as a bridge rectifier, it is not as popular because the centre-tapped transformer is more expensive and more bulky.

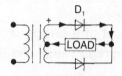
▶ On the first half cycle of the sine wave the top of the secondary winding becomes positive with respect to the centre tap. Current flows through the diode D_1 and through the load resistor to the centre tap. This produces a positive half cycle across the load.

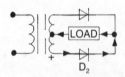
▶ On the second half cycle of the sine wave the bottom of the secondary winding becomes positive with respect to the centre tap. Current flows through D_2 and the load resistor to the centre tap. This produces a second positive half cycle because the current flows in the same direction through the load. This is called a *biphase circuit*.

EXERCISES

▶ Build the circuit shown in the diagram.

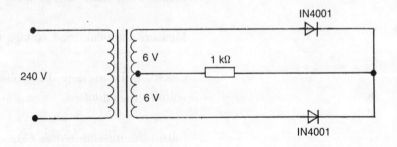

Check the d.c. output using a voltmeter.

Check the peak d.c. output using an oscilloscope.

Calculate the average value of the d.c. voltage and compare it with the voltmeter reading.

Fill in the missing words (5).

It is possible to produce a .... wave output using only two diodes but a .... transformer is needed which is more .... and more ....

■ ■

Filters

▶ The voltage output at this stage is a succession of positive half cycles. This now has to be modified to produce a smoother d.c. A *filter circuit* is used to do this. The simplest form of filter circuit is a *capacitor filter*.

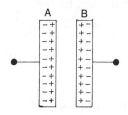

▶ An uncharged capacitor has an equal number of negative electrons on each plate of the capacitor.

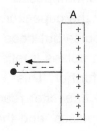

▶ When a d.c. supply is connected to the capacitor with its positive terminal connected to plate A and its negative terminal to plate B, the electrons on plate A will be attracted to the positive terminal of the supply, leaving an excess of positive charges on that plate.

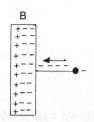

▶ The negative terminal will supply electrons which are attracted to plate B, until it is saturated with electrons, making it negatively charged. The capacitor is now charged and all current flow stops.

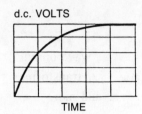

▶ Charging the capacitor does not happen instantaneously and a time delay is involved. The delay depends on the *charging rate* and the capacitance value of the capacitor.

▶ As the charging of the capacitor is not instantaneous this gives the effect of capacitance delaying the voltage as it tries to rise.

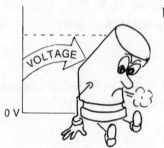

▶ There is also a time delay when the capacitor is discharging. This depends on the discharge rate and the capacitance value of the capacitor. The effect is for capacitance to delay the voltage as it tries to fall.

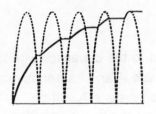

▶ The broken line in the diagram shows the full wave d.c. output from the bridge rectifier and superimposed on it is the effect of time delays produced by the capacitor. The capacitor partly charges on the first pulse but does not rise as sharply as the pulse. Between pulses it falls back a little and the next pulse charges it further. This continues until the voltage across the capacitor reaches the peak value of the supply voltage pulses, and the current flow to the capacitor ceases.

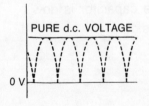

▶ Without a load on the capacitor, it will not discharge, so that once it has reached its maximum charge the voltage will neither increase or decrease but hold a steady level, giving a pure *d.c.* output voltage, shown opposite as a straight line.

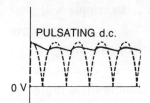

▶ When a load is connected the capacitor charges during each pulse and discharges during the interval between each pulse by supplying current to the load.

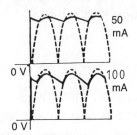

▶ When the load current is small, the *ripple* will be small but as the load current increases then the ripple also increases. A larger capacitor has a longer time delay, so that the ripple is again reduced. However, the larger the capacitor, the greater the size and the greater the cost.

E X E R C I S E S ▶ Build the circuit shown in the diagram.

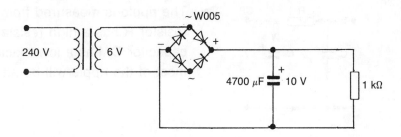

Measure the d.c. output using a voltmeter.

Remove the capacitor and measure the output.

Note the waveforms both with and without the capacitor and sketch them.

Reduce the load resistor to 500 Ω and note the effect on the waveform.

Fill in the missing words (6).

The waveform with the capacitor in circuit is much .... The output voltage is .... because the capacitor charges up to the .... value of the waveform and not the .... value.

POWER SUPPLY UNITS

If the load current is .... the ripple will
.... and a larger .... will have
to be used to achieve the same amount of smoothing.

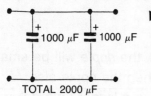

▶ Two capacitors placed in parallel doubles the plate area and so doubles the capacitance, but capacitors alone will not always smooth the output enough and a more sophisticated filter is required.

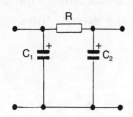

▶ The R–C_2 network is called an *RC filter*. Ignoring C_1 which has previously been discussed R and C_2 are seen to be in series with each other forming a voltage divider.

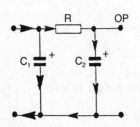

▶ The ripple is measured from peak to peak (P × P). The resistor R has a high resistance to the ripple while the capacitor C_2 has a low reactance to the ripple so that most of the ripple will be attenuated by the resistor.

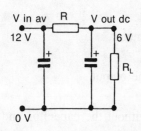

▶ For light load requirements, this is an excellent filter, but the resistor R forms a d.c. voltage divider with the resistance of the load, R_L. For example, when the load resistance is the same value as R, the d.c. output voltage will be reduced to half the average value of the input voltage.

EXERCISES

▶ Build the circuit shown in the diagram.

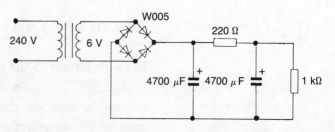

Note the ripple and compare it with a 9400 μF single capacitor filter.

Measure the output voltage and compare it with the single capacitor filter.

Reduce the load resistor to 500 Ω and measure the output voltage for both types of filter.

Fill in the missing words (7).

The RC filter is .... than the .... filter for smoothing out the a.c. ripple, but the d.c. output voltage is .... If the load .... is heavy or varying the .... filter is not suitable due to the voltage drop across the series resistor.

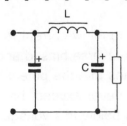

▶ For heavier loads an *LC filter* is more suitable, as a choke has a high reactance to a.c. but a low resistance to d.c. Therefore it reduces the ripple with little output loss.

▶ The disadvantage of a choke is that it is large, heavy and has a relatively high cost, so that they are rarely used.

Voltage stabilisation

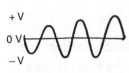

▶ The r.m.s. value of the mains supply varying, results in the PSU d.c. output voltage varying. Also, as the d.c. load current taken from the PSU increases, the d.c. output voltage drops. To counteract these effects, *voltage stabilisation* is used.

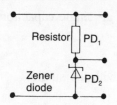

▶ The simplest form of *voltage stabiliser* or *regulator* is the *Zener diode*. The diagram shows a Zener diode in series with a resistor across the d.c. supply. Two potential differences occur, one across the resistor (PD_1) and the other across the Zener diode (PD_2).

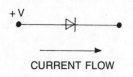

▶ When a Zener diode is connected in a forward biased direction (i.e. the anode is connected to the supply and the cathode to the output), it behaves like a normal diode and passes current.

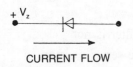

▶ However, the Zener diode has a special characteristic in that it passes current when reversed biased, provided the voltage reaches a certain specified value V_z.

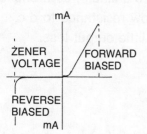

▶ The Zener diode is specially designed to allow current to flow at a pre-determined voltage when reverse biased and survive the resultant current. This means that the potential difference across the Zener diode can never exceed the pre-determined value and can only fall below this voltage if it does not receive a minimum current requirement.

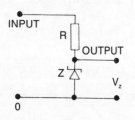

▶ Since the voltage across the Zener diode is constant it is possible to use the potential difference across it as the constant output voltage (V_z). This means that if the input voltage rises and the Zener voltage remains the same the additional voltage drop is across R.

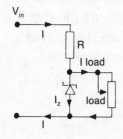

▶ When the load current increases the output voltage will not be affected until the load current becomes so great that the Zener diode does not receive the necessary current for effective operation.

INVESTIGATING ELECTRONICS

RESISTANCE

$$R = \frac{V_{in} - V_z}{I_z}$$

▶ The value of the series resistor, *R*, can be calculated as shown. The voltage drop across *R* can be found by subtracting the Zener voltage V_z from the input voltage V_{in}. By dividing this with the typical Zener current found in data sheets which also flows through *R* the value of *R* can be obtained (Ohm's law).

EXERCISES

▶ Calculate the value of *R* in the circuit shown.

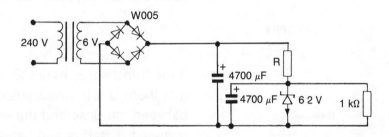

Build the circuit shown in the diagram.

Measure the output voltage.

Measure the load current.

Replace the load resistor with a 500 Ω resistor and measure the output voltage and the load current.

Replace the load resistor with a 220 Ω resistor and measure the output voltage and the load current.

Determine the voltage drop across *R*.

Fill in the missing words (8).

When the load .... varies the output .... varies unless a .... is used.
The more .... drawn by the .... ,
and the bigger the voltage drop across the ....

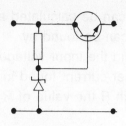

▶ A big disadvantage of using the Zener voltage to supply the load R_L, is that both R and R_L are in series. R is therefore limiting the load current as well as the Zener current. This can be overcome in an improved voltage stabiliser or *voltage regulator*, using a transistor.

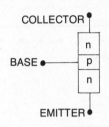

▶ The construction of the transistor is very similar to that of the diode with an extra n region and a very narrow central p region, each region having its own connection and named base, collector and emitter.

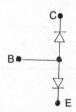

▶ If the transistor is re-drawn as two diodes, it can be seen that there is a low resistance or forward bias region between the base and the emitter and a high resistance or reversed biased region between the collector and the base.

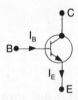

▶ If the base voltage is made higher than that of the emitter then a current will flow from the base to the emitter. In practice this forward bias voltage must be around 0.7 volts.

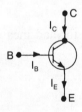

▶ When a base-to-emitter current flows the high-resistance region between the collector and the emitter is broken down and current flows from the supply into the collector and out of the emitter. The emitter current will be the sum of the base current and the collector current, that is, $I_E = I_B + I_C$.

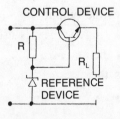

▶ As the voltage across the reference *Zener diode* is constant, the voltage at the base of the *control transistor* is constant. This means that the transistor is switched on to the same extent irrespective of the supply voltage and the resistor R has no effect upon the output voltage. As the transistor needs 0.7 V between the base and the emitter to switch it on the output voltage is 0.7 V less than the Zener voltage.

EXERCISES ▶ Build the circuit shown in the diagram.

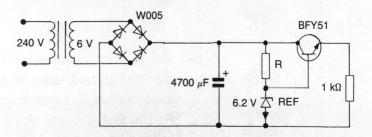

Measure the output voltage.

Measure the load current.

Change the load resistance to 500 Ω and measure the output voltage and current.

Change the load resistance to 220 Ω and measure the output voltage and current.

Fill in the missing words (9).

This .... is an improvement because most of the .... current does not flow through the ...., so that the .... at the output varies very little. Only when the .... current becomes excessive will the .... stop working.

Excess-current protection

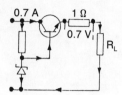

▶ To provide an *excess-current protection* of, say, 700 mA, a resistor of about 1 Ω is placed in series with R_L, as shown. The voltage drop across this resistor is 1 Ω × 0.7 A, that is, 0.7 V.

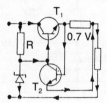

▶ The protection transistor T_2 is connected as shown. When the load current reaches about 700 mA, the 0.7 V at its base causes T_2 to conduct, this current passing through R. The voltage drop across R increases and reduces the voltage at the base of the control transistor T_1. The control transistor tends to pass less current, which protects it from excess current damage.

E X E R C I S E S ▶ Build the circuit shown in the diagram.

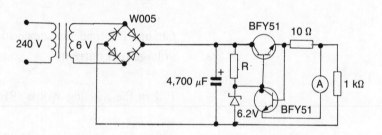

Record the reading on the ammeter.

Reduce the load resistor to 100 Ω and record the ammeter reading.

Fill in the missing words (10).

If the load current is .... the control .... could be damaged. Excess current can be .... by using a second .... which only .... when excess .... flows through the control ....

98   INVESTIGATING ELECTRONICS

Commercial voltage regulators

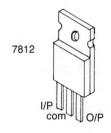

▶ The 78 series fixed voltage regulator is a positive-voltage *integrated circuit regulator*. Two additional figures denote the output voltage (e.g. 7812 is a 12 V regulator). It is rated at 1 A provided that it is mounted on a suitable heat sink. The leads are marked input, common and output.

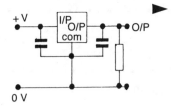

▶ The regulator incorporates a control device, a reference device and overload protection. To enable correct operation it must be connected using the capacitors and resistor as shown in the diagram.

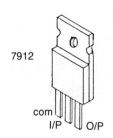

▶ A 79 series negative fixed voltage regulator is also manufactured and is complementary to the 78 series. The most obvious difference is that the leads are in a different order. The 7912 regulator is a negative 12 V regulator.

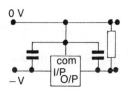

▶ The circuit opposite shows the 79 series voltage regulator with the capacitors and resistor required for its correct operation.

- -

E X E R C I S E S ▶ Build the circuit shown in the diagram.

POWER SUPPLY UNITS

Measure the load voltage.

Reduce the load resistor to 220 Ω and measure the load voltage.

Fill in the missing words (11).

The 78 series fixed .... is an integrated circuit comprising a ...., a .... and .... protection incorporating some external components.

. . . .

E X E R C I S E S ▶ Build the circuit shown in the diagram.

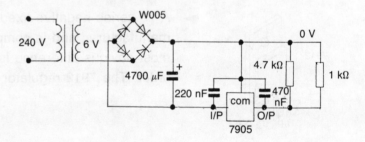

Measure the output voltage.

Replace the load resistor with a 220 Ω resistor and measure the output voltage.

Fill in the missing words (12).

The 79 series .... is a .... version of the 78 series .... The last two figures denote the .... voltage.

. . . .

EXERCISES ▶ Build the circuit shown in the diagram.

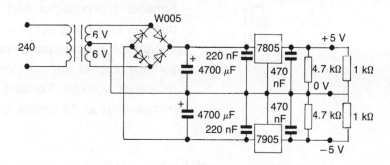

Measure both the output voltages, noting the polarities of each.

Fill in the missing words (13).

It is possible to build a .... with both .... and .... outputs, but a .... transformer must be used to give a .... voltage output.

Variable-output voltage regulators

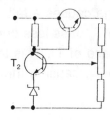

▶ Voltage regulation can be further improved and made variable by the use of a *feedback network* of resistors and a second transistor known as the *comparison transistor* T_2.

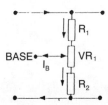

▶ The diagram shows the feedback network. This is a voltage divider connected across the output. Current will flow through the network from positive to negative. If the base of the transistor is connected to the wiper of the variable resistor (VR_1) some current (I_B) will flow into the base/emitter junction and switch on the transistor.

POWER SUPPLY UNITS 101

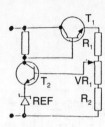

▶ When the variable resistor (VR$_1$) is varied until the base current of T$_2$ is at the maximum, then the resistance between the collector and emitter, and hence the voltage drop, is at a minimum, i.e. T$_2$ is turned on. The base voltage of T$_1$ is equal to the sum of the collector/emitter voltage drop of the comparison transistor plus the reference voltage. Consequently the output of the control transistor is at its lowest level.

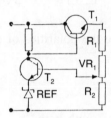

▶ When the variable resistor (VR$_1$) is varied until the base current of T$_2$ is at its minimum, then the resistance between the collector and emitter, and hence the voltage drop, is at its maximum, i.e. T$_2$ is turned off. The base voltage of T$_1$ is at its highest, consequently the output at the emitter of T$_1$ is at its highest.

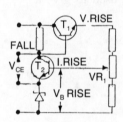

▶ With the variable resistor (VR$_1$) set to give a particular voltage output, any rise in this voltage causes a rise in the base voltage of T$_2$, tending to make T$_2$ pass more current, thus reducing the voltage at the base of T$_1$. This tends to reduce the current flow through T$_1$, causing the output voltage to fall back to its correct value.

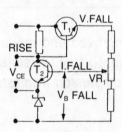

▶ With the variable resistor (VR$_1$) set to give a particular voltage output, any fall in this voltage causes a fall in the base voltage of T$_2$, tending to make T$_2$ pass less current, thus increasing the voltage at the base of T$_1$. This tends to increase the current flow through T$_1$, causing the output voltage to rise to its correct value.

EXERCISES ▶ Build the circuit shown in the diagram.

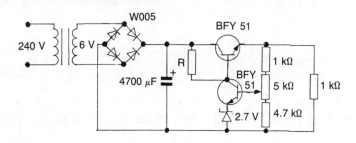

Set the potentiometer to give maximum output voltage and record this voltage.

Note the effect on the output voltage of slowly decreasing the potentiometer resistance to a minimum.

Measure the minimum output voltage.

Fill in the missing words (14).

The output voltage can be .... by decreasing the voltage at the base of the .... transistor. The effect of this is to .... the voltage at the base of the .... transistor, causing it to pass more current and thus .... the output voltage.

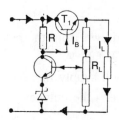

▶ If a very heavy load I_L is experienced it will reduce the output voltage beyond the control of the voltage regulator. This is because the base current of T_1 (I_B) is limited by the value of resistor R.

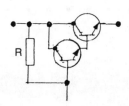

▶ This can be overcome by using *Darlington pair* transistors as the control device in place of T_1.

POWER SUPPLY UNITS

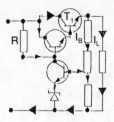

▶ The control transistor of the pair T_1 does not have its base current (I_B) limited by the value of R and more current (I_C) can be supplied to the load before voltage control is lost.

EXERCISES ▶ Build the circuit shown in the diagram.

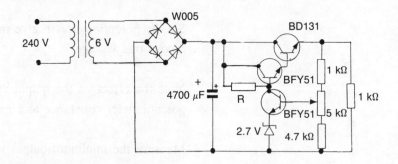

Set the output voltage to 6 V.

Reduce the load resistor to 100 Ω and measure the output voltage.

Replace the Darlington pair with a single control transistor and measure the output voltage.

Fill in the missing words (15).

If a .... load current is drawn from the power supply the output .... will fall. A .... will supply more .... to the load which raises the output ....

Matrix board construction

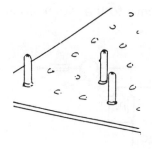

▶ *Matrix board* is an insulating board perforated with a uniform pattern of holes all over its surface. Pins can be pushed into the holes on which to make the circuit connections. The distance between the pins will depend upon the components that are used. The advantage of matrix board is that the circuit can be easily traced as the wires and the components are on the same side of the board.

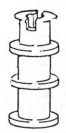

▶ In addition to pins, wires and components can also be attached to terminal posts. The terminal post shown is called a *turret post* because of the castellated appearance at the top.

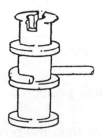

▶ Wire is wrapped around the post to form a walking-stick shape. The contour of the connection wire should be plainly visible after soldering. Two or three connections are needed on each post and all these connections must be separate. Care must be taken not to overtighten leads, and components must be central.

▶ In a turret post terminal, the cable form leads can be fed up through the hollow centre of the terminal and wrapped around the top groove. However, this type of termination will be difficult to remove.

Practical exercise: TTL power supply unit

Cut a piece of perforated fibre board to the dimensions shown.

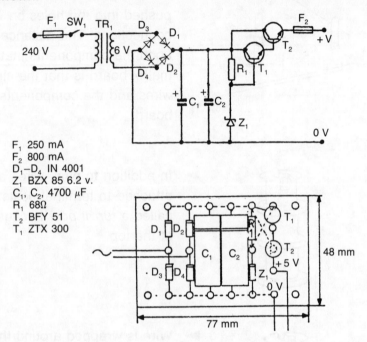

F$_1$ 250 mA
F$_2$ 800 mA
D$_1$–D$_4$ IN 4001
Z$_1$ BZX 85 6.2 v.
C$_1$, C$_2$ 4700 μF
R$_1$ 68Ω
T$_2$ BFY 51
T$_1$ ZTX 300

Rivet turret posts in the relevant positions.

Wrap and solder the components into position.

Feed the wires up through the posts from the bottom and solder on the top turret.

Fill in the missing words (16).

Matrix board is a .... insulation board. The components are jointed on .... which are pushed into the board. The advantage of matrix board is that the .... and the .... are on the .... side of the board and .... the circuit is very easy.

106 INVESTIGATING ELECTRONICS

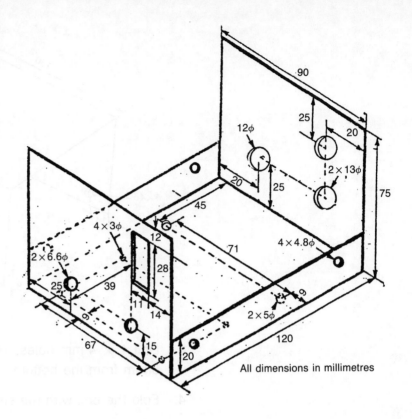

All dimensions in millimetres

THE CHASSIS

1. Cut a piece of steel or aluminium alloy 130 mm × 270 mm.

2. Mark out the shape of the complete chassis with a scriber and rule and cut to shape.

3. Centre punch mark and drill all the holes.

4. Punch the oblong hole with a metal punch.

5. Fold the box so that the oblong hole is on the right-hand side.

THE COVER

1. Cut a piece of aluminium alloy 242 mm × 130 mm.

2. Mark the folding lines 75 mm from each end.

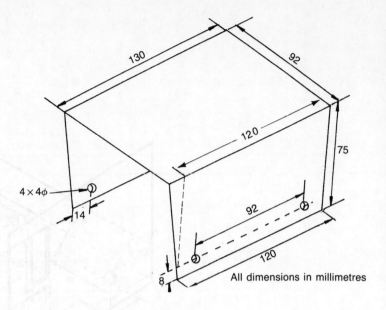

3. Drill 4 × 4 mm holes, 14 mm from the edge and 8 mm from the bottom.

4. Fold the box with the stelvatite surface on the outside.

THE ASSEMBLY

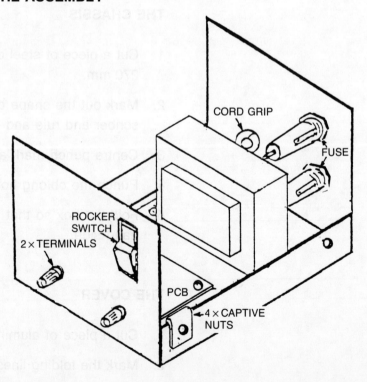

1. Label the front panel, if required.
2. Mount the switch and terminals to the front of the box.
3. Fit the transformer into the box.
4. Mount two fuse holders into the rear panel.
5. Feed the three-core mains cable through the back panel.
6. Stanchion the PCB inside the box using 25 mm insulated pillars.
7. Make all the electrical interconnections inside the box and tie in cable form. Sleeve all of the board joints.
8. Fit the lid using four captive nuts to accept the lid-fixing screws.

COMPONENTS LIST

- 1 TRANSFORMER
- 4 JUNCTION DIODES IN 4001
- 1 ZENER DIODE 6.2 V
- 1 TRANSISTOR BFY 51
- 1 TRANSISTOR ZTX 300
- 2 ELECTROLYTIC CAPACITORS 4700 μF
- 1 68 Ω RESISTOR
- 1 ROCKER SWITCH (SPST)
- 2 20 mm FUSE HOLDERS
- 1 800 mA FUSE 20 mm
- 1 250 mA FUSE 20 mm
- 1 INSULATED TERMINAL (BLACK)
- 1 INSULATED TERMINAL (RED)
- 18 TURRET TAGS
- 1 PERFORATED BOARD (77 mm × 48 mm)
- 1 m, 3-CORE, ROUND FLEX
- 1 CABLE GRIP
- 1 PLUG TOP
- 8 CAPTIVE NUTS
- 4 SELF-TAPPING SCREWS
- 4 RUBBER FEET

8   SELF-TAPPING SCREWS
4   PLASTIC SPACERS (25.4 mm)
2   SHROUDED RECEPTACLES
4   2BA SCREWS
4   2BA WASHERS
4   2BA NUTS

Connection to mains supply

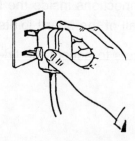

▶ One of the jobs often asked of an electrician is to connect a *plug top*.

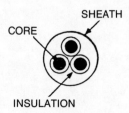

▶ The type of cable used for this purpose is called three-core *insulated and sheathed*. It has colour-coded insulation on each of the copper cores for electrical safety and an overall sheath for protection against damage.

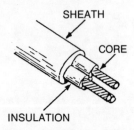

▶ The sheath and the insulation are usually PVC (polyvinylchloride). PVC is reasonably cheap and resistant to corrosion. The cable strands are made from copper as this is the best cost effective conductor of electricity.

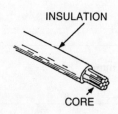

▶ *Multi-strand cable* is used where flexible wiring is required. The cable size may be given as 16/0.2 mm, this means that the cable has sixteen strands and each strand has a diameter of 0.2 mm.

0.5 mm = 2.5 A
0.75 mm = 6.0 A
1.0 mm = 10 A

▶ 3-core 16/0.2 mm has a c.s.a. (cross-sectional area), of 0.5 mm^2.

0.5 mm cable has a current rating of 2.5 A. If the appliance to be connected to the plug has a current rating higher than this, a larger cable size must be chosen.

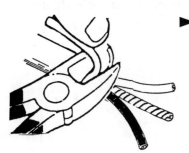

▶ When preparing the cable it is first necessary to remove the protective sheath. This is done using side cutters and cutting down its length, taking care not to damage the insulation.

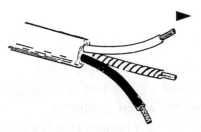

▶ When the required length has been cut the sheath is pulled back and trimmed off.

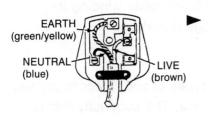

▶ After the sheath has been removed the colour-coded insulation is seen. The tips of the insulation are removed for connection purposes.

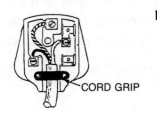

EARTH (green/yellow)
NEUTRAL (blue)
LIVE (brown)

▶ When connecting the plug, the brown live wire is connected to the fuse holder on the right of the plug, the blue neutral to the left and the green and yellow earth wire to the largest pin at the top of the plug.

▶ It is important that the cord-grip is clamped tightly to the outer sheath of the cable, to remove the strain from the connections.

POWER SUPPLY UNITS

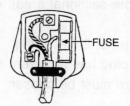

▶ The fuse is now replaced but it is necessary to choose a fuse which is rated lower than the rating of the cable being used. Be sure to replace the cover before using the plug.

E X E R C I S E

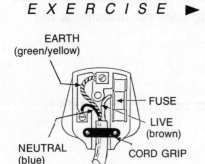

▶ Strip the cable to the required length for connecting the cable to the terminals. Strip the inner insulation so that the insulation is close to, but not under, the clamping screw and that every strand of wire is connected. Ensure the wires go to the correct pins. Tighten the cable clamp firmly on to the cable sheath and replace the cover.

Testing insulation

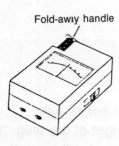

Fold-away handle

▶ It is not safe to switch on a mains circuit without first testing it. The type of meter used for this test is a *megger tester*. The megger is a generator which produces electricity at about 500 V by turning a handle to drive it. It is important not to touch the leads while winding the handle, due to the high voltage produced.

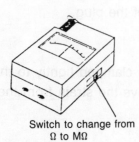

Switch to change from Ω to MΩ

▶ The megger is used for two types of test, a *continuity test* and an *insulation resistance test*. The continuity test is performed using the ohms range of the meter and the insulation resistance using the megohms range. A switch on the side of the case is used to select either the ohms range or the megohm range.

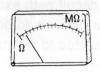

▶ The meter also has two scales. An ohm scale reads from left to right and a megohm scale reads from right to left.

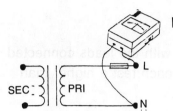

▶ When checking a circuit's continuity, switch to the ohm range, connect the leads to the live and the neutral wires and turn the handle of the megger.

▶ Good continuity means that there is a low resistance path round the circuit in which the current flows easily. The resistance reading should be approximately the resistance value of the load (in this case the transformer primary resistance plus 1 Ω).

▶ If the meter reads higher than this, it could indicate a high resistance joint which will become hot in use and may cause an electrical fire.

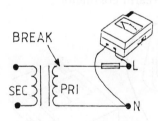

▶ If the meter reads hard over to the right, this indicates a break in the circuit.

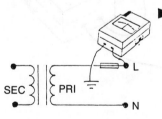

▶ When checking the insulation resistance the switch must be set to megohms and one lead connected to the live conductor and the other to the earth connection, in this case, the metal-work associated with the transformer.

▶ When the handle is turned, the meter should read infinite resistance, shown by the symbol ∞. This means that there is no connection between the live wire and the earth.

► If the meter reads lower than 1 MΩ, a circuit path exists between the live wires and earth and a fuse would blow if the circuit was switched on when connected to the mains supply.

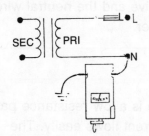
► The test must now be repeated with the leads connected between neutral and earth. For each test a higher than 1 MΩ reading must be obtained.

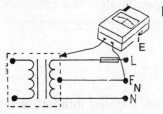
► If the circuit is housed in a metal case an earth continuity test must be performed on the case. Set the megger to the ohms range and connect the leads to the earth terminal and the metal case to be tested. When the handle is turned, as with all continuity readings, the meter should read less than 1 Ω.

• •

EXERCISES ► Using a megger, test the circuit for earth continuity.

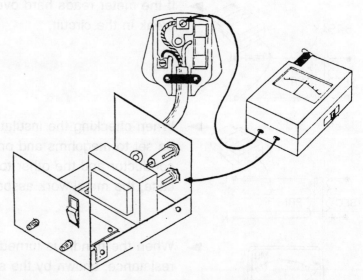

Test the insulation resistance between the live conductor and earth.

INVESTIGATING ELECTRONICS

Test the insulation resistance between the live conductor and neutral.

Test the insulation resistance between the neutral conductor and earth.

Fill in the missing words (17).

A megger is a hand-powered .... and is used to measure the .... between two points. If the megger is set to Ω a .... reading is being taken but if it is set to MΩ an .... test is being performed.

· ·

Self-assessment questions

1. A transformer changes;

(a) a.c. to d.c.,
(b) d.c. to a.c.,
(c) high voltage to low voltage,
(d) half wave to full wave.

2. A centre-tapped transformer has;

(a) one secondary winding,
(b) two secondary windings,
(c) three secondary windings,
(d) four secondary windings.

3. A single-diode rectifier changes;

(a) a.c. to pulsating d.c.,
(b) a.c. to smooth d.c.,
(c) d.c. to pulsating a.c.,
(d) d.c. to smooth a.c.

4. A bridge rectifier is constructed from;

(a) one diode,

(b) two diodes,
(c) three diodes,
(d) four diodes.

5. A biphase rectifier circuit provides

(a) quarter-wave rectification,
(b) half-wave rectification,
(c) full-wave rectification,
(d) double-wave rectification.

6. A capacitor filter;

(a) filters out the d.c. component,
(b) reduces the ripple,
(c) reduces the output voltage,
(d) changes a.c. to d.c.

7. An RC filter is most suitable for;

(a) heavy load currents,
(b) varying load currents,
(c) light load currents,
(d) all types of load current.

8. The Zener diode is most suitable as a;

(a) control device,
(b) rectifier device,
(c) overload protection device,
(d) refrence voltage device.

9. In a voltage regulator, the control device is a;

(a) diode,
(b) capacitor,
(c) Zener diode,
(d) transistor.

10. The 7812 integrated circuit regulator is a;

(a) 12 V positive fixed regulator,
(b) 12 V positive variable regulator,
(c) 12 V negative fixed regulator,

(d) 12 V negative variable regulator.

11. A half-wave filtered power supply peaks at 12 volts. The useable d.c. will be;

(a) 12 volts,
(b) 8.4 volts,
(c) 7.6 volts,
(d) 3.8 volts.

12. It is an advantage to use a Darlington control device if;

(a) the a.c. component is high,
(b) the current is light,
(c) the current is heavy,
(d) the rectification is half wave.

13. r.m.s. is;

(a) the effective d.c. voltge,
(b) the average d.c. voltage,
(c) the effective a.c. voltage,
(d) the average a.c. voltage.

14. The r.m.s. value of an 8.5 volt sine wave is;

(a) 8.5 volts,
(b) 6.5 volts,
(c) 6 volts,
(d) 5 volts.

15. To transform a fixed regulator into a variable regulator we must add;

(a) a feedback network and a reference device,
(b) a comparison transistor and a reference device,
(c) a control transistor and a feedback network,
(d) a comparison transistor and a feedback network.

Answers to the self-assessment questions

1. (c); 2. (b); 3. (a); 4. (d); 5. (c); 6. (b); 7. (c):

8. (d); 9. (d); 10. (a); 11. (d); 12. (c); 13. (d); 14. (c); 15. (d).

Missing words

(1) Voltmeter, r.m.s., peak, peak, root mean square, effective.
(2) Centre tapped, positive, centre tap, negative, centre tap, double.
(3) Average, effective, average, half.
(4) Halve, transformer, effective, four.
(5) Full, centre tapped, bulky, costly.
(6) Smoother, higher, peak, average, high, increase, capacitor.
(7) Better, capacitor, lower, current, RC.
(8) Current, voltage, voltage regulator, current, load, d.c. output.
(9) Voltage regulator, load, series resistor, voltage, load, voltage regulator.
(10) Excessive, transistor, shunted away, transistor, switches, current, transistor.
(11) Voltage regulator, control device, reference device, overload.
(12) Voltage regulator, negative, voltage regulator, output.
(13) Power supply, positive, negative, centre tapped, zero.
(14) Increased, comparison, increase, control, raise.
(15) Heavy, voltage, Darlington pair, current, voltage.
(16) Perforated, posts, components, wires, same, tracing.
(17) Generator, resistance, continuity, insulation.

INDEX

ammeter 2
ampere 2
amplifiers
   differential 70
   feed back 67
   inverting 74, 75
   non-inverting 77, 78
   operational 73
   stabilised 67
   voltage 59
amplitude 62
amplitude distortion 61
anode (diode) 9
anode (SCR) 48
armature 41
attenuator 62
average voltage 85, 87

back e.m.f. 44
base (transistor) 23
batteries 81
bell 51
biasing 59
biphase rectifier 88
bit (soldering iron) 28
bridge rectifier 86, 87
button (preset resistor) 16

cable
   insulated and sheathed 110
   multistrand 110
capacitor
   electrolytic 21
   polarised 21
   unpolarised 21
capacitor filter 89
cartridge fuse 4
cathode (diode) 19
cathode ray tube 63
cathode (SCR) 48
charging rate 90
circuit board
   copper clad 52
   printed 52
circuit diagram 2
coil 82
cold detector 38
collector (transistor) 23
colour code (resistor) 9
comparison transistor 101
contacts
   change over 43
   normally closed 42, 43
   normally open 43
   retaining 46

continuity test 112
continuity tester 40
control gate (SCR) 48
control transistor 96
coupling characteristics 62, 76
current
   direct 1
   electrical 1
   limiting resistor 20

Darlington pair transistor 103, 104
d.c. voltage 90
detector
   cold 38
   darkness 36
   heat 39, 45
   level 39, 47
   light 37, 45
   two-stage 47
differential amplifier 70
digital multimeter 11
diode 85
   forward biased 19
   junction 19
   light emitting 20
   reverse biased 19
   Zener 94
direct current 1
double-pole double-throw switch 8
double-pole single-throw switch 6
duel in line 74

electrical current 1
electrolytic capacitor 21
electromagnet 41
emitter (transistor) 23
etching 52
excess-current protection 98

fault finding 24
feedback bias 67
feedback network 101
filament 1
filter
   capacitor 89, 90, 91
   LC 93
   RC 93
flow chart 25
flux (soldering) 28
forward-biased 19
forward current transfer ratio 60
frequency 62, 64
full wave rectification 86, 88
fuse 4
   cartridge 4

gate (SCR) 48

half-wave rectification 85
heat detector 39
heat shunt 29
hFE 60, 66
Hertz 64

insulated and sheathed cable 110
insulation test 112
integrated circuit operational amplifiers 73, 74
integrated circuit regulator 99
inverting amplifier 74, 75

LC filter 93
level detector 39, 47
light-dependent resistor 18, 36
light-detector 37, 45
light-emitting diode 20
long-tailed pair 70

magnetic field 81, 82
matrix board 105
megger tester 112
milliamperes 2
multimeter (digital) 11
multistrand cable 110

negative terminal 2
non-inverting amplifier 77, 78
normally closed (contacts) 42, 43
normally open (contacts) 43
NPN (transistor) 23, 49

offset null 75
ohmmeter 10
Ohm's law 12
operational amplifier 73
oscilloscope 63
overdriven 61

parallel 3
peak voltage 3, 85
plug top 109, 111
p-n junction 85
PNP transistor 23, 49
polarised capacitor 21
positive terminal 2
potentiometer 17
power rating 27
power supply unit 81
preset resistor 16
  button 16
  skeleton 16
primary winding 82
printed circuit board 52
protection
  excess current 98

quiescent point 67

R-C filter 92
rectification
  full wave 85
  half wave 86
rectifier (silicon controlled) 48
reference Zener diode 96
regulator
  integrated circuit 99, 100
  variable voltage 101, 102
  voltage 94, 96
relay 41, 42
resistance 1
resistor 9
  colour code 9
  current limiting 20
  light-dependent 18
  preset 16
  tolerance 10
  variable 15
resistor (light-dependent) 36
retaining contacts 46
reverse biased (diode) 19
rheostat 15
r.m.s. voltage 83

secondary winding 82
sensitivity 37
series connection 2, 13
signal generator 62
silicon controlled rectifier 48
skeleton preset resistor 16
solder 28
  flux 28
soldering 27
  iron 27
  iron bit 27
stabilisation
  voltage 93
stabilised amplifier 67
step-down transformer 83
stripboard
  cutter 29
switch 5
  double-pole double-throw 8
  double-pole single-throw 6
  single-pole double-throw 7
  single-pole single-throw 5

tapped winding 84
terminal
  negative 2
  positive 2
terminal block 68
test
  continuity 112
  insulation 112
tester
  continuity 40

megger 112
thermistor 17, 38
thyristor 48
timebase 64
tinning 27, 49
tolerance (resistor) 10
transformer 81
  step-down 83
transistor 22, 96
  base 23
  collector 23
  comparison 101
  control 96
  Darlington pair 103
  emitter 23
  NPN 23, 49
  PNP 23
turret post 105
two-stage (detector) 47

unpolarised capacitor 21

variable resistor 15

variable voltage regulator 101, 102
virtual earth 75
voltage 2
  average 85
  d.c. 90
  peak 83, 85
  regulator 94, 96
  r.m.s. 83
  stabilisation 93
voltage amplifier 59
voltage drop 14
voltmeter 3

wattage 2
winding
  primary 82
  secondary 82
  tapped 84
wiper 15

Zener diode 94
  reference diode 96